Up from the Underworld

UP FROM THE UNDERWORLD

COALMINERS AND COMMUNITY IN WONTHAGGI

1909–1968

Andrew Reeves

© Copyright 2011
All rights reserved. Apart from any uses permitted by
Australia's Copyright Act 1968, no part of this book may be
reproduced by any process without prior written permission
from the copyright owners. Inquiries should be directed to the
publisher.

Monash University Publishing
Building 4, Monash University
Clayton, Victoria 3800, Australia
www.publishing.monash.edu

This book is available online at
www.publishing.monash.edu/books/ufu.html

ISBN: 978-0-9806512-6-3 (pb)
ISBN: 978-0-9806512-7-0 (web)

Design

Les Thomas

Cover images

Front cover painting: Noel Counihan, *Miners working in wet conditions* (1945). National Gallery of Australia, Canberra, purchased 1974.

Background cover photo: Aerial view of Twenty Shaft only hours after the explosion. Wonthaggi Historical Society Collection.

Frontispiece image

The banner of the Federation's Wonthaggi Branch is carried for the first time at the 1958 Melbourne May Day demonstration. Wonthaggi Historical Society Collection.

Closing image

Noel Counihan, *In the narrow seam*, linocut, 1947.
Private Collection.

Printer

Griffin Press

'The pit is to the village what the tree is to the dryad. When the pit dies, the village dies too; when the pit is ill, the village groans.'

Aneurin Bevan, quoted in Michael Foot,
Aneurin Bevan, Vol 1, London, 1962, p. 34.

Contents

Acknowledgments — *ix*
Acronyms — *x*
Preface — *xi*
Introduction — *xiii*

Chapter 1
Building a future: The early years, 1909–1914 — 1

Chapter 2
Fragile prosperity: Consolidation and expansion, 1914–1929 — 22

Chapter 3
Up from the underworld, 1929–1934 — 53

Chapter 4
Political recovery and economic decline, 1934–1939 — 79

Chapter 5
Contraction and contradiction, 1939–1945 — 116

Chapter 6
'Back the Miners' Programme': Isolation and confrontation, 1945–1949 — 131

Chapter 7
'When the last wheel finally turns…' 1949–1968 — 146

Bibliography — *153*
Index — *161*

Acknowledgments

This book draws heavily on research I undertook in 1973–76 for a Master of Arts degree in History from La Trobe University. At that time I offered heartfelt thanks to many people who had assisted in my research and I thank them once again. Alas, many have since died, but my debt to them is in no way diminished. In no particular order, thanks are due to Nancy Rankine, Bill and Nancy Stirton, Harry Brydon, Agnes and Wattie Doig, Alan Opie, Harry Bell, John McKenzie, Joe and Lyn Chambers, John Montgomery, and all of Wonthaggi; to Mark Richmond, Frank Strahan, Liz Gallagher, Kelvin Rowley, Grant Evans, Janeen O'Connell and the late John Knight of the Victorian Department of Mines. Particular thanks go to my supervisor, the late Dr Peter Cook, an outstanding historian and teacher.

In preparing this current work others have provided generous support. Above all, thanks are due to Luke van der Meulan of the Victorian Mining and Energy Division of the CFMEU and to Paddy Gorman of that union's National Mining Office. I have drawn on Peter Cochrane's 1973 thesis, 'The Wonthaggi Coal Strike, 1934', for his treatment of that strike and on Dr Charles Fahey's invaluable 1987 monograph, *The Wonthaggi State Coal Mine*, commissioned by the Victorian Department of Conservation, Forests and Lands. I must thank the members of the Wonthaggi Historical Society, particularly Irene Williams and Sam Gatto, not only for their insights into Wonthaggi's history, but also for their generosity in making available many of the photographs used in this book, and Mick Counihan and the other trustees of the Noel Counihan Estate for their generous permission to reproduce a number of Noel Counihan's Wonthaggi paintings and prints. Thanks also to Nina Divich and the team at Monash University Publishing, to Neil Newitt for his typically expert assistance with photography, to Tess Brady for discussions on writing and authorship, to Stan Anson for proofing and editing, to my sons Keir and David for their constant advice and criticism, and to Cora Trevarthen, for without her enthusiasm and imagination this book would have remained locked in a thesis.

Acronyms

ACSEF	Australian Coal and Shale Employees Federation
AEU	Amalgamated Engineers Union
AFULE	Australian Federated Union of Locomotive Engineers
ALP	Australian Labor Party
ARU	Australian Railways Union
AWU	Australian Workers Union
CFMEU	Construction, Forestry, Mining and Energy Union
CPA	Communist Party of Australia
FEDFA	Federated Engine Drivers Federation of Australia
FOSU	Friends of the Soviet Union
ILP	Independent Labour Party
IWW	Industrial Workers of the World
MM	[Militant] Minority Movement
OBU	One Big Union
PLC	Political Labor Council
PLL	Political Labor League
RSL	Returned and Services League of Australia
THC	Trades Hall Council
TLC	Trades and Labour Council
VCAWF	Victorian Council Against War and Fascism
VCMA	Victorian Coal Miners Association
VSP	Victorian Socialist Party
WIUA	Workers Industrial Union of Australia
WIIU	Workers International Industrial Union
WUWU	Wonthaggi Unemployed Workers Union

Preface

This book addresses a simple question: how was it that a small mining community in Victoria with a strong union, on the margin of Australian coal mining, was able to exert an influence on that industry out of all proportion to its size and numbers for nearly 60 years?

In Victoria black-coal mining started later and finished sooner than in Tasmania, but no Tasmanian community or mine workforce exerted the same influence over the industry that Wonthaggi's did. Indeed, for many years the Wonthaggi miners formed a sub-group within a Victorian/Tasmanian branch of the Australian Coal and Shale Employees Federation. Similarly, although Wonthaggi's mineworkers represented a small fraction of the union's national membership, their role in the union's recovery following the depression of the 1930s proved as strategic as that played by the larger districts in New South Wales. In particular, it was Wonthaggi and its success in the 1934 strike that provided the Federation with the blueprint for community mobilisation that was to be used successfully for the following decade.

At different stages of its history the State Coal Mine, together with its workforce, represented a national benchmark for mine development, the successful operation of a state enterprise and the exercise of political influence within the coal industry. Even so, the history of Wonthaggi as an influential mining town remains a conundrum. For nearly 60 years the State Mine proved an economic aberration, working the thin, broken seams of a coalfield that would have been considered uneconomical in the opinion of most Australian mine owners. A small regional mining community that won a national reputation, Wonthaggi came to be admired by many and disliked by some. Such contradictions can be resolved, in part, by recognising that for Wonthaggi's miners their place within a national union was as important as their regional sensibility, and that the solution to local issues depended, to a significant degree, upon national answers. The union, by whichever name it was known between 1909 and 1968, proved central to the self-perception and identity of the town and its place in wider industrial and social issues. This book seeks to tell something of the history of this union and its community.

Introduction

When George Macartney, a former spruiker for the Victorian Railways, published his ambitiously titled *Victorian Coal Consumers' and Investors' Guide* in the winter months of 1894, the colony of Victoria had just begun to show tentative signs of recovery from nearly four years of depression. For 20 pages, Macartney extolls the virtues of this neglected industry, one that he describes as 'the most important industry, next to that of gold, which has taken place in Victoria'.[1]

The coal industry he describes proved as modest as his prospectus. Its publication followed hard on the heels of a Royal Commission in 1889 that had led to the first direct government support for the industry, and although by 1894 Macartney could list 16 black-coal companies active in the Gippsland region, only three companies reported that they were actually in production. Nine were 'under construction' or still proving reserves, others provided no further information. Even so, in that year the Gippsland mines produced 171,000 tons of coal, a figure barely exceeded over the next decade. The rate of attrition for such speculative companies proved to be high. And speculative they were. Although the land boom and the depression that followed had been difficult, Macartney's pamphlet suggests that a number of the participants remained keen to try their hand again. No fewer than seven politicians were listed on the boards of these companies, some holding multiple memberships. The two politicians acknowledged in Macartney's preface, F L Outtrim and J H McColl, had both been involved in financial swindles during the previous decade, and McColl, in particular, had a history of interest in Gippsland commercial affairs.

Macartney's slim pamphlet introduced many of the themes that would bedevil the industry until the establishment of the Wonthaggi State Coal Mine in 1909, and a few that would persist well beyond that time. These included the operational problems facing an industry characterised by small under-capitalised companies: low productivity and intermittent production, political interference and manipulation, and, above all, a sense of conspiracy that the potential of a nascent Victorian industry was being squeezed by the 'Coal Masters of New South Wales'.

---

1   George Macartney. 1894. *The Victorian Coal Consumers' and Investors' Guide*. Melbourne: Chas Troedel & Co.

Macartney's optimism stood at odds with the anarchic history of the industry in Victoria. The existence of coal outcrops across a swathe of Gippsland had been well known since the earliest days of settlement, and from the mid-nineteenth century a number of unsuccessful attempts had been made to create an industry to supply the rapidly growing Victorian market. By the 1890s, however, the industry's focus had moved from the coastal deposits around Cape Paterson and Kilcunda to the hill towns of Korumburra, Outtrim and Jumbunna (the phase of mining that Macartney promoted), but here, too, the industry could not escape the twin evils of underinvestment and high cost of production. The industry carried high expectations and some considered these years 'boom years'. If such a boom existed, driven particularly by coal contracts with the Victorian Railways, it was short-lived. While the number of miners working the Victorian coal mines grew to nearly 1300 by early in the twentieth century, these miners suffered from chronic job insecurity, short-time, primitive working conditions and pressure on wages. By 1905 all but three of the companies Macartney had promoted a decade previously had failed, and these remaining companies – Coal Creek, the Jumbunna Coal Company and the Outtrim-Howitt Company – now controlled 85 per cent of declining local production.[2]

By 1905, the industrial and political landscape had also changed dramatically. The Victorian industry's crisis in production and profitability climaxed in 1903, when the larger proprietary companies sought to reduce production costs by introducing a new agreement that effectively cut wage rates by up to 30 per cent. Employer refusal to conciliate and union resistance to wage cuts resulted in a 70-week lockout from all company pits in Gippsland. Employer refusal to negotiate on any basis other than unconditional acceptance of their original proposals ultimately served to reduce union resistance, notwithstanding financial support for the mining unionists from across Australia. Under increasing pressure, the Gippsland-based Victorian Coal Miners Association (VCMA) began to fragment, and by May 1904, when organised resistance finally ceased, the union had been virtually destroyed.

For some, the union's demise offered an opportunity not to be missed. In addressing the 1906 Victorian Royal Commission on the Coal Industry, Frederick McCoy from the Coal Creek Company

> admitted that immediately after the strike certain men were refused employment, but that no distinction was now made unless the men are

---

[2] See Victorian Royal Commission. 1906. *Report of the Victorian Royal Commission on the Coal Industry*. Appendix A. Melbourne: Government Printer: 51.

well-known agitators… these men he believed were not at present in the state.[3]

Unsurprisingly, others saw matters in a different light. As one unionist told a follow-up Parliamentary Select Committee:

> The only thing I think I had the sack for was because [I] was in the union… I left of my own accord because I could not make a living… I was sent to a bad place [in the mine] because I was a unionist.[4]

This selective weeding out of union members created a generation of 'wild geese': ex-Victorian miners later to be found in Western Australia and New Zealand, on Broken Hill's 'line of lode' and in Illawarra coal mines. The ideologies of industrial militancy to which these miners were exposed would return to Wonthaggi a generation later.

In the meantime, unionists remaining in Victoria sought to rebuild the VCMA under the twin protections of arbitration and parliamentary support. Despite determined employer opposition, the union successfully applied in September 1907 to be registered under the Commonwealth's *Conciliation and Arbitration Act 1904*. But the VCMA faced the more immediate problem of recognition in the Victorian mines themselves. At Coal Creek, for example, the VCMA had to compete not only with the breakaway South Gippsland Miners Association but also with the company-sponsored Victorian Colliery Owners and Employees Federation. 'Free' (non-union) labourers worked side by side with VCMA members, with the union incapable of enforcing its authority. Political warfare was waged below ground: unionists would be sent to unproductive sections of the mines, while sympathetic wheelers (responsible for delivering skips – or coal trucks – to miners and then moving filled skips to the surface) retaliated by depriving free miners of sufficient skips to load their coal.

For the coal companies, the victory in 1904 had proved pyrrhic. Neither production nor profitability recovered. Of the remaining proprietary companies, the Coal Creek Company ceased production in 1907. Coal imports from New South Wales soared as local production collapsed. By 1909, local production had fallen to half the level of a decade previously, and the poor condition of the surviving companies only served to emphasise the industry's stagnation and inability to supply even its local market.

---

3   ibid: xvi.
4   See *Report of the Select Committee on the Coal Mining Industry*. 1907. Melbourne: Government Printer: xviii.

A series of Royal Commissions and Parliamentary Select Committee Inquiries between 1904 and 1907 sought to identify solutions to the twin dilemmas of declining local production and Victorian dependency on coal from the trouble-prone mining districts of New South Wales. Some argued for a greater emphasis on exploration, and as a consequence, a geological team led by Stanley Hunter discovered the rich Powlett River seams in 1908. The Labor Party, supported by a number of Liberal members, argued that these deposits should be reserved for state use. Some went further, arguing that the Powlett seams should be developed as a state-owned coal mine. These supporters included mine inspectors from the local industry, members of the Labor Party – growing in political importance – and, significantly, Liberal MP and Minister for Mines Peter McBride. The emerging Labor parliamentary leader George Prendergast voiced typical sentiments when he argued in July 1908:

> I understand that the diamond drill working down in the Powlett River country has discovered a seam of coal that is the best discovered up to the present time in this State… If the Mines Department is going to allow private speculation at present when we want coal for our railways, it is doing an unwise thing. I want the Premier to say that no speculator shall be allowed to get hold of the land, but the State will reserve it for its own use.[5]

A radical few argued that such a mine should be accompanied by a model state town that would include planned housing for workers and adequate provision of civic amenities, providing security and dignity to those so often denied such social benefits. There the issue may have languished but for developments in New South Wales, where a bitter protracted dispute known as the Peter Bowling Strike provided the necessary catalyst to transform parliamentary debate into government action.

While the VCMA actively supported the NSW strikers, and sought to use its own federal registration to bring about arbitration court intervention, it would be the state government that would take the most significant initiative. By mid 1909 the Victorian Railways had been reduced to using wood to fuel some locomotives, so great was their dependence upon (now unobtainable) Newcastle coal, and this crisis led to manufacturing dislocation and mass industrial layoffs that threatened to grind the railway system to a halt. The Bowling Strike re-emphasised the economic necessity of constant, reliable coal supplies.

---

5   *Victorian Parliamentary Debates (VPD)*: 1908 Session. 7 July 1908: 28.

Confronted by the inability of local collieries to increase production, the government adopted the argument for a state-run mine on the Powlett Plains. An emergency Cabinet meeting in early November 1909 approved a recommendation by McBride to establish a state coal mine. On 11 November preliminary work on the Powlett mine commenced under the direction of Stanley Hunter, with 50 goldminers recruited from Rutherglen in the state's north-east. By Christmas, a temporary supply line had been opened, using bullock teams to carry coal 11 miles to Inverloch for shipment to Melbourne by sea.[6]

McBride's vision reshaped not only Victoria's coal industry but state politics as well. His advocacy established Australia's first state-owned coal mine. It conjured expectations of not only a model town – Wonthaggi – but also of a secure future for mineworkers, expectations that attracted hundreds of miners who migrated from the hill towns of Gippsland's Strezlecki Ranges to the windswept, desolate Powlett Plains within months of the State Mine's establishment. Wonthaggi's State Coal Mine carried a burden of expectation from its first days, and the subsequent history of the mine and its workforce reflects a determination to achieve both the security and the social opportunity previously denied to miners in Victoria.

---

6 For a general description of the mine's initial months, see George Broome. 1910. 'State Coal Mines'. *Annual Report of the Secretary for Mines for the Year 1910*. Melbourne: Government Printer: 139–141.

Chapter 1

Building a future

The early years, 1909–1914

Ten months after the first coal shipments were dragged by bullock to the coast, *The Criterion*, one of Wonthaggi's newly established newspapers, surveyed progress. It drew a picture of a frontier shanty settlement straggling eastward from a temporary, government-run tent town that abutted the State Mine's operations to occupy the site surveyed for the permanent township:

> Looking toward the west, in the foreground at the foot of the hill is the Government township, built on the site of a gently sloping rise – a unique collection of buildings of all sizes, shapes and material and in varying stages of construction, mingling with the primitive surrounds of the bush. In parts, the land has been cleared and street formed, but in others, dead trees fallen and dead trees waiting to be felled are on the same allotments as buildings or shops that are almost complete.[1]

While the permanent town described by *The Criterion* rapidly grew on the north–south ridge dividing the Powlett's major coal basins, a population estimated at more than 2000 lived in a government-maintained tent town close to the mining operations. This was a community short on amenities and distractions: no women or children were permitted to live in the tents – the few who came originally to the site lived during these early months at the small village of Dalyston, four miles to the north-west – while the limited amenities included a temporary theatre, travelling vaudeville acts, boxing troupes, wood-chopping contests and a few grocers who had arrived with Stanley Hunter's original goldminers.

---

1 *Criterion*, 10 September 1910.

After 15 months the mine's general manager, George Broome, could boast that 'the quick development at the State Coal Mine constitutes a world record in coal mining' and list an impressive tally of developmental works, infrastructure and modern mining practices to support his claim, including one of the first electrically driven hauling engines installed in an Australian mine, and remote-control endless-rope haulages in two of the most productive shafts.[2] Born in England in 1866, Broome came to Wonthaggi with extensive mining experience in Britain, New Zealand and Canada. His experience in superintending the establishment of the Stockton Mine on New Zealand's rugged West Coast, where he introduced endless-rope haulages, electrical coal cutters, fans and pumps, stood Wonthaggi's new state mine and its miners in good stead. Broome came to the State Mine with a reputation of having never lost a single day's work to industrial disputes. While Wonthaggi's industrially active workforce soon changed that, Broome retained his own preference for a unionised workforce at the mine, successfully arguing to government the benefits of dealing with 'collective bodies' rather than individuals.

Even so, the realities of social life in Wonthaggi compared badly with the mechanical amenity of the State Mine itself. Despite the state government's rhetoric about a new start for the industry, many miners feared a continuation of the cycle of decline and primitive conditions that had marked the previous decade. As one anonymous miner, 'Shovelman', told a local paper:

> They tell you to come along here, we find you plenty of work, we build you a nice house, we let you buy it cheap and only charge a little rent and we make [a] nice township for you and treat you working men nice and proper. Well, now, sometimes we get work and sometimes we get good pay until they think we make too much money and then they cut the [contract] price down. They build a lot of houses and put in weatherboards too green and hardwood, not pine, and next winter when they have got through the hot summer, you just come along and poke your nose between each board.[3]

These early months were also a time of periodic mass departures of miners and their families as erratic production and interrupted employment underpinned a climate of economic uncertainty, in turn exacerbating poor

---

2   George Broome. 1912. 'The State Coal Mine'. *Coalfields and Collieries* in Australia. F D Power (ed.). Melbourne: Crichley Parker: 283.
3   'Shovelman', *Criterion*, 30 July 1910.

living conditions in the newly established community. Health inspectors' reports of inadequate and unsanitary drainage, of lax building regulations and of boarding houses providing rudimentary accommodation for single miners compared to the conditions prescribed for English workhouses reinforced Shovelman's complaints. Everyone needed a bit of luck. For recently arrived Coal Creek miner John (Jack) McVicars (who in 1913 would commence a 32-year stint as the Federation's Victorian secretary), luck came in the form of unexpected accommodation: 'On arrival in Wonthaggi [in 1912] our party of six were fortunate in securing from Mr Mick Cosgrove a large galvanised room previously used to store horse fodder, in which we batched for some time, and later brought our families to Wonthaggi.'[4] Others boarded in Wonthaggi, returning to family in Melbourne or elsewhere between sets of shifts. The absence of any commercial liquor licences – a legacy of state control of the original tent town – led to the popularity of 'blue pigs' (shanty stills) hidden in the surrounding scrub. The forced pace of development, harsh living conditions and lack of social amenities were represented in one widespread complaint: 'that few, if any, of the promises held out by the government as a bait to induce settlement have been fulfilled'.[5]

What were these promises, and who had made them? The most potent, in the short term at least, was the state government's promise of a 'state town' at Wonthaggi and the expectations that this generated among the growing local population. Official rhetoric lent plausibility to popular belief: the new township would be laid out on model lines and the state government would construct 100 houses to house mineworkers and their families, at an expense of A£20,000. Indeed, mineworkers 'were to have every consideration that a beneficial state can afford them'.[6] But while Victoria's *Coal Mines Regulation Act 1909* undoubtedly led to vastly improved working conditions underground, there were no equivalent social gains in the new, planned community, and the language of a promised state town would be jettisoned soon after the Bowling emergency had passed. Prospects for a state town proved illusory. The government's support for a state mine had been based on a pragmatic desire to protect the state's growing manufacturing economy by 'providing financial stimulus to an essential, yet ailing and previously unprofitable industry'. Under such circumstances, a model state-run town clearly represented unnecessary additional cost.

---

4   Untitled manuscript. Autobiographical memoir of John (Jack) McVicars, n.d., Wonthaggi Historical Society.
5   Quoted in the *Powlett Express*, 30 September 1911.
6   Mr Toucher MLA, in *VPD*: 1909 Second Session, 1 December 1909: 26–36.

Figure 1. Long wall mining, Twenty Shaft, State Coal Mine, c. 1920s
Wonthaggi Historical Society Collection

Figure 2. Drilling the coal seam preparatory to shot firing, Powlett River coalfields, c. 1910
Wonthaggi Historical Society Collection

Figure 3. Cutting coal by hand at the coal face, possibly in the State Mine's Western Area, c. 1930s
Wonthaggi Historical Society Collection

Figure 4. 'Tent town', close to the newly established State Mine pit heads, c. 1909–10
Wonthaggi Historical Society Collection

Figure 5. Mineworkers queuing at the State Coal Mine pay office, c. 1912
Wonthaggi Historical Society Collection

Figure 6. Bullock teams and horses carrying wood for the construction of the State Mine, 1910
Tent town is visible in front of the tree line at the rear of the mine workings.
Wonthaggi Historical Society Collection

Anticipation of a state town at Wonthaggi soon turned into a more acute community question – would the real potential of the Powlett River coalfields ever be realised? The growing fear that it would not stemmed from the decision in 1911 to transfer control of the State Mine from the Victorian Mines Department to the Victorian Railway Commissioners, thereby effectively endorsing the commissioners' view of the mine as a 'strategic reserve'. Despite the necessity to maintain regular production, the mine's operations would henceforth be dictated by two over-riding conditions: the need to deter excessive price exploitation by NSW producers; and the mine's consequent value as a permanent state-owned coal reserve, assuring regular supplies to the Railways and other government departments should coal from interstate sources be disrupted by industrial action or any other cause. This emphasis on potential value rather than what the community saw as real, if unrealised, value was popularly seen as a straightjacket for both the mine and the dependent community, and one difficult to escape. In the town, some campaigned for an amendment to the State Mine's charter to allow unhindered coal sales to public consumers, a solution seen as a sure way of realising Wonthaggi's future as 'the Newcastle of the South'.

Within 12 months of its establishment, the State Mine employed nearly 900 miners, a figure that grew rapidly to 1200 by 1912 and fluctuated around that mark for the next decade. It took many skills to support the miners actually working at the coal face: engine-drivers, fitters and other engineers, blacksmiths and their assistants to keep picks sharp and equipment in trim, carpenters and power-house engineers, and stablemen and horse breakers to tend the small army of pit ponies required to haul coal underground. A hierarchy of miners had quickly been established. When first employed in the mine it was traditional for a young man to work in the brace building, where coal was brought to the surface and sorted into round (premium) coal or 'slack' (broken) coal. In 1919, the mine employed 24 fourteen-year olds, 30 fifteen-year olds and 16 eighteen-year olds in this work, as a preliminary to going underground. Brace boys could graduate to working as 'clippers', responsible for attaching skips to the endless-rope haulages, or as 'wheelers', who drove coal skips from the coal face to the endless ropes for transport to the surface. A good wheeler might be promoted to contract mining in their mid-to-late twenties. Others found employment as roadsmen, laying rails to the coal face and ensuring that the 'roads', or tunnels, were kept clear. Contract miners represented the aristocrats of the mine's workforce, and all other workers laboured to keep them as productive as possible.

The mine's original Rutherglen goldminers had quickly been joined by recruits from the declining goldfields of Central Victoria, from Walhalla – a Gippsland gold town set deep in the Great Dividing Range – and, more significantly, from the troubled coal centres of Korumburra, Outtrim and Jumbunna in the Strezlecki Ranges to the north of Wonthaggi. It would be this latter group who would bring the VCMA to the State Mine. On Christmas Eve of 1909, a hastily convened general meeting held in tent-town's biograph marquee voted to establish a union. Before the meeting broke up, official positions had been filled on a provisional basis, with many key positions occupied by names familiar to Gippsland coal miners: Frank Murphy as secretary, Ern Yardley as treasurer, and Matthew McMahon as a committeeman. Members agreed to reconvene a month later, following the Christmas shutdown, to formalise the union's structure and constitution. By the time that meeting took place, on 23 January 1910, a vigorous debate on the name and affiliation of the new union had erupted. Given the preponderance of Gippsland miners who had flocked to Wonthaggi, most unionists had assumed, logically enough, that the union formed at the State Mine would automatically become a branch of the VCMA; but Murphy, as secretary, begged to differ. Never short of an opinion or an idea, Murphy sought to put his stamp on the new union. Arguing that Wonthaggi miners needed a separate union to recognise the State Mine's 'unique circumstances', Murphy pushed for an Australian miners' union formed along the lines of the Australian Workers Union (AWU) and constitutionally separate from the VCMA. It is not clear how widely Murphy had canvassed this idea, but he had certainly misjudged his audience. Speaker after speaker spoke up for the VCMA. Matthew McMahon fought to 'stay with the VCMA', arguing that no new body would be recognised, whatever links it sought to establish with mining unionists elsewhere. In an argument that would echo down the years of Wonthaggi's history as a mining community, Bert Holland appealed to fellow members to not separate Wonthaggi from the 25,000 other coal unionists organised nationally. Such sentiments carried the day, and the meeting agreed that the names of all unionists at the State Mine would be sent to the general secretary of the VCMA for registration. The meeting also appointed a Rules Committee and nominated organising stewards to actively recruit among the growing workforce.

What lay behind Murphy's argument? Certainly, mining unions across eastern Australia were in a state of flux during these years and different organisational models had been advocated. By his reference to the AWU, Murphy also demonstrated an acute understanding of union growth in

Australia; by 1910 the AWU stood on the cusp of a series of amalgamations and recruitment drives that would sweep goldminers and other hard-rock miners into its ranks and stamp it as a union with unrivalled power and influence. But the miners seeking work at the State Mine were looking for a new start, not a new union. Wonthaggi's proximity to the smaller, older coal towns of Gippsland, the continuing migration of miners from these centres to the State Mine and the effort that had been put into rebuilding the VCMA all ensured that the VCMA never lacked a voice or majority support among the mine's workforce.[7]

Despite a hiccup in mid February when it emerged that the members' names had not been forwarded in time for Wonthaggi unionists to vote in a VCMA district election (a difficulty accepted as an oversight), a formal VCMA branch was in full operation by May. The union's centre of gravity inevitably shifted to Wonthaggi as its population expanded, a fact confirmed by the relocation of Earn Yardley's hairdressing saloon from Outtrim to Wonthaggi. A former miner himself, Yardley served as the first Branch treasurer and his saloon as the union's de facto office on the Powlett fields, with the names of non-unionists publicly displayed in the saloon's front window during the union's earliest membership drives. Yet the fact that a hairdresser, however sympathetic, acted as union treasurer reflects the initial instability of the union at Wonthaggi and the need for officials immune from potential victimisation. Matters soon consolidated, though, and by 1914 the position of treasurer would be absorbed into the fulltime secretary's position.

While the wave of strikes that had washed over the Australian coal industry since the turn of the century had left the majority of Wonthaggi miners with a distinct preference for political reform over 'direct action' as their means of driving social improvement, it proved to be unsuccessful strike action that ultimately cemented the position of the union at the State Mine. In this, the union was aided by none other than the mine's general manager, George Broome. A number of the former goldminers mistrusted both unionism and strike action and the VCMA faced a challenge to convince them otherwise. But in June 1910 Broome elected to replace the miners' standard 10 shilling daily rate that had applied since the mine's initial developmental period with contract rates based on a sliding scale, a proposal that threatened to reduce the wages of some miners. Broome demanded unconditional acceptance of his non-negotiable terms. In response, the miners stopped work.

---

7   For accounts of the earliest union meetings see the *Powlett Express*, 7 January 1910; 28 January 1910; 18 February 1910; 29 April 1910.

Figure 7. The first political meeting at the Powlett River, in support of the Political Labour Council (Labor Party), January 1910
Wonthaggi Historical Society Collection

Figure 8. The first union Vigilance (or organising) Committee, Powlett River coalfields, January 1910
Frank Murphy stands at top left with Matthew McMahon on his immediate left. Harry Webb, Committee Secretary, sits at Murphy's feet.
Wonthaggi Historical Society Collection

Figure 9. The first miners' union committee, Powlett River coalfields, 1910
Frank Murphy, the first secretary, is second from the left in the middle row; Matthew McMahon, President, sits on Murphy's immediate left. Second from the right in the same row is Jack Goldsmith, a long-time militant in Wonthaggi union ranks.
Wonthaggi Historical Society Collection

Figure 10. Members of the union's Victorian District Committee, c. 1914–15
Jack McVicars sits third from the left in the front row.
Wonthaggi Historical Society Collection

Although the strike proved both short and unsuccessful, it did serve to consolidate union authority and isolate those 'few crawlers who, having obeyed the behests of their employers with cringing servility, have refused to join or encourage the unions'.[8] Miners reluctantly acquiesced to Broome's demands following assurances that wage cuts were not intended, but sporadic industrial action continued as other grievances continued to fester, particularly the union's right to control the 'cavil', and a parallel campaign to end the mine's third daily shift. British in origin, the cavil represented a union-controlled system of work allocation for miners employed on contract, or piece rate. In practice it took the form of a quarterly random ballot for 'bords', or workplaces, underground that prevented, for example, 'non-unionists from Outtrim [being] given preference of easy work… and generally "coddled" by the management'.[9] Conducted every three months, the State Mine cavil covered the entire mine – every working pit and development area. Miners balloted in teams of four and much hung on the result – a good result meant extra earnings for the next three months, while a bad result could mean unpaid bills and increased debt. In 1910 the union successfully insisted on retaining the cavilling system for work allocation, but the issue of the third shift proved much more volatile. A three-shift system boosted the miners' employment security. It undoubtedly provided economies of scale in terms of mine operations but also resulted in a chronic state of over-production, far exceeding railway requirements. But rather than abandon the 'dog watch' (night shift), as union policy demanded, Broome sought to dampen production by means of periodic lay-offs, arguing that the abolition of the third shift could only lead to permanent retrenchments. In March 1912 the union brought matters to a head by refusing to allow its members to work the dog watch. Broome bowed to this decision but enforced one of his own – he immediately stood down 400 mineworkers and dropped the contract rate for coal from 3 shillings to 2 shillings and 9 pence, claiming that the extra three pence per ton had, in fact, been an allowance for night work. Staggered by the magnitude of this response, the VCMA threatened further strike action. Broome stood his ground, and on 27 March all mining ceased at the State Mine.

The union did not approach a protracted stoppage with equanimity, and made repeated efforts to negotiate a settlement to this six-week strike. Broome rejected numerous union concessions and proposals and

---

8   *Criterion*, 29 April 1911.
9   *Sentinel*, 11 June 1910.

also deflected the VCMA's efforts to have the issues in dispute referred to arbitration. Donations of food and cash from local businesses supplemented the union's own funds, but resources proved limited. In early May the union sued for peace, agreeing on 11 May to a 12-month agreement that confirmed 2 shillings and 9 pence per ton as the prevailing contract price for coal.

Two grievances lingered. Miners believed that their determination to end the dog watch had saved the State Mine's economic position, but that they had been forced to carry the full financial cost of that decision. Similarly, they had accepted the contract rate reduction because of economic necessity, not economic justice. Long-time miners recalled that 'it was an old Outtrim trick to first starve the men and then compel them to accept whatever was offered'.[10] The new start that the State Mine had seemed to offer in its earliest months now appeared as far off as ever and relations between mineworkers and management deteriorated accordingly. The fact that a shortage of developmental capital for the State Mine had had a knock-on effect on the local economy during 1912 did not help either. The failed six-week strike left the union in a quandary. On one hand, the union could not avoid becoming embroiled in the myriad disputes at the coal face that were a fact of life in this sort of mining; on the other, it determined to avoid further lengthy disputes, to the extent of entering into an accord with Broome in early 1913 to curtail independent action by State Mine wheelers, agreeing to a code designed to curb unauthorised stoppages and expose those who ignored this code to management discipline.

Even so, by 1912 the union could consider its position with a degree of satisfaction. The organising efforts of early 1910 had proved successful. Union membership reached high levels at the State Mine, with the exception of salaried and clerical staff. Basic union principles such as the use of the cavil to allocate work on a non-preferential basis had been successfully defended, giving the union a significant presence in the mine's day-to-day operations. But perhaps the best evidence of the speed with which the union exerted its influence at the State Mine can be found in a paragraph from the *Powlett Express* in late April 1910, less than six months after the mine's establishment:

> Eight Hours Day was a general holiday at the State Mine camp, all work being suspended. A sports meeting was held in the afternoon, with foot running and log chopping contests taking place. There was a large attendance present from all parts.[11]

---

10   *Sentinel*, 23 July 1910.
11   *Powlett Express*, 29 April 1910.

Although originally gazetted as a public holiday by the Victorian Government as early as 1879, 30 years later workers and their unions, particularly in regional centres, still had to 'claim' the day. The fact that the *Powlett Express* could so matter-of-factly report Wonthaggi's first Eight Hours Day celebrations reflects the degree to which union claims and union assumptions had infiltrated the culture of the new town.

In the years immediately preceding the First World War, a revitalised VCMA played a crucial role in the creation of a national coalminers' union. Wonthaggi's short history and the diversity of the State Mine's workforce aided its creation. In contrast, the local unions that organised miners on the NSW coalfields were hamstrung by entrenched regional allegiance, accentuated by the failure of earlier industrial campaigns. Wonthaggi miners were never particularly susceptible to such parochialism and, if anything, at this time, related more to trans-Tasman influence. It had been to New Zealand that many black-listed Victorian coalminers had moved following the failure of the 1903–04 strike, and a decade later a number had achieved positions of influence within the radical wing of New Zealand labour, particularly the militant Federation of Labor – the 'Red Fed'. Bob Semple, later to be New Zealand's first Labour Minister for Public Works and Transport, mined at Korumburra in 1903 and, as the VCMA's Branch president there, had been a natural target for employers. Influential in New Zealand mining politics as well as the Red Fed, Semple regularly visited family in Wonthaggi before the First World War, constantly urging closer co-operation and organisation among Australasian unionists. For example, during a visit in May 1912, he called for coalition between miners and wharfies to 'more efficiently conduct class wars and secur[e] industrial solidarity in… industrial struggles', and on his return to New Zealand announced that the miners of NSW, Victoria and New Zealand had been 'linked up'.[12]

The immediate aims of Victoria's miners proved more prosaic than Semple's grand vision, preferring, in the words of the Branch president Tom Burley, 'to see a federation of coal miners first'. Such a preference reflected the prevailing belief among VCMA members that wider amalgamation and political effort offered the best formula for winning improved wages and conditions. 'The union', its officials constantly reiterated, 'should organise first and strike afterwards'. In March 1913, VCMA members belatedly endorsed at least one element of Frank Murphy's earlier vision, voting 675 to 81 to alter the union's title to the Australasian Coal Miners Association

---

12 *Criterion*, 4 May 1912; 28 September 1912.

(ACMA). At the same time, renewed agitation in NSW coal districts over wage rates similarly promoted a further attempt at national union: 'The time has arrived when it is essential in the interests of the coal-mining community that the respective districts federate in one organisation for mutual protection'.[13]

Yet Powlett miners had not been invited to the first of these unity conferences, a problem not remedied until July 1915 when representatives of NSW, Queensland, Victorian and Tasmanian coalminers met as one union, forming the Australian Coal and Shale Employees Federation (ACSEF). The emphasis on district autonomy and on co-operation between districts, as opposed to northern NSW district domination, reflected the success of Victorian arguments and provided the new union with the flexibility to survive its early years. As the rules of the new union advised: 'Be practical at all times. Leave theorising to spring poets'.[14]

A similar preference for consensus and political action marked the union's activities in Wonthaggi. At a meeting in tent town's centre of community life, the biograph marquee, in early January 1910, a large roll-up of miners heard a Political Labor League (PLL, or Labor Party) organiser, Miss Harriet Powell, urge them to join. A branch was formed the following week, spurred on by the prospect of a federal election barely a month away. This new PLL branch soon expanded beyond its initial mineworker base, growing into a broader community organisation with a bias for local issues. Although the union, by virtue of its State Mine membership, had the weight of numbers to dictate the course of local politics and social activities, the VCMA much preferred to collaborate with other unions and sympathetic local businessmen. This form of collaboration owed something to mutual economic interests but also reflected the town's changing political landscape, especially after the state government formally established Wonthaggi as a borough in 1911. As Harry Bloustein, a local tailor and prominent member of the Labor Party, argued:

> The miners are here to get a living and the businessmen [to make] a living through them. No attempt at class distinction should be made, as the miner and the businessman wanted to work together.[15]

---

13 For Burley's preference see *Criterion*, 20 November 1913; for officials' advice to members see *Criterion*, 17 December 1910; for the decision on union federation see Edgar Ross. 1970. *A History of the Miners Federation of Australia*. Sydney: Miners Federation of Australia: 257.
14 Australasian Coal and Shale Employees Association. 1915. *Rules of the Australasian Coal and Shale Employees Association*. Sydney: ACSEF: 3.
15 *Sentinel*, 9 July 1910.

And work together they did, ensuring that the Wonthaggi Labor Party Branch soon assumed the role of arbiter in local politics: 'it is confidently expected that an increase in labour members at council tables would be all the better for the community'.[16] The Labor allegiance of many local businessmen also reflected the opportunities for social mobility offered in a new and expanding community, and the early ranks of local businessmen were soon swelled by ex-unionists. J W Webb, the VCMA's first Wonthaggi president, soon emerged as the proprietor of the Caledonian Hotel, financed by John Wren, while Matthew McMahon, another early president, retired from union affairs to become the local representative for the Monobel Powder Company, which supplied the State Mine.

The confirmation of Wonthaggi's borough status in mid 1911 provided a further opportunity for this union–business partnership. When, after a crushing victory at the first municipal election, nine successful Labor candidates met as one of Australia's first all-Labor municipal councils, six businessmen and three working miners took their seats. Tension existed; not everyone agreed that Wonthaggi should be a self-governing borough – the memory of the promise of a state town still proved potent. But Labor businessmen remained firm, and unionists conceded the point. For these businessmen, Wonthaggi's 'great future' depended, in part, on the translation of their property leases into freehold titles, and despite Labor Party policy to the contrary, this was the position that the union–business alliance advocated. Union-supported attempts to implement an early finish for municipal employees were similarly frustrated. But by and large the mutual benefits of such political domination over-rode such differences, and an agreed emphasis on practical politics rather than ideological grandstanding reinforced its effectiveness: 'Labor policy [should] embrace a comprehensive socialism, without making everyone adhere to the dull uniformity of always wearing a red tie or persistently Shibboleth mouthing.'[17] Or as Murphy, representing the union, observed, the importance of local businessmen in the party provided proof that 'Labor was not a bigoted, class-conscious movement'.[18]

As the Wonthaggi Political Labor Council (PLC) increasingly developed as a successful electoral machine, the Council served as a forum for Labor principles and policy, with differences in interpreting Labor policy often resolved at Council meetings. But beyond the boundary of

---

16   *Criterion*, 1 April 1911.
17   *Criterion*, 20 May 1911.
18   *Criterion*, 8 July 1911.

municipal politics, community initiative remained firmly in the hands of the miners' union. Tied to the rhythms of municipal process, the PLC could not match the socially conscious activism of the union, one continually reinvigorated by the effect of pit-head disputes as well as the inherent dangers of mining itself. Improved social amenities remained a priority for the union. It would be union, rather than PLC, initiative that would create the Workingmen's Club, the Co-operative Dispensary and the Co-operative Store. Significantly, all occupied high-profile positions on Graham Street, Wonthaggi's principal commercial thoroughfare. Likewise, the local hospital was established in 1910 following urgent union representation. Miners contributed the majority of the hospital's funds (70 per cent in 1911, for example) and provided a clear majority on the hospital's committee of management. The union also organised Wonthaggi's coal-carting service for the home delivery of coal.

Not all union members supported the union's embrace of the PLC, and after August 1911 the union harboured a socialist ginger group. Although never large, this group did contain a number of experienced miners who lent it a degree of credibility. Foremost among them was Jack Goldsmith, an experienced miner with an acerbic tongue, together with Jack Burns and possibly Archie Nelson, later a Labor councillor. The Wonthaggi group affiliated with the Melbourne-based Victorian Socialist Party (VSP), a well-established organisation that could trace its origins back to the Australian visit of English socialist Tom Mann some years earlier.

The VSP's attitude toward the Labor Party dictated the actions of Wonthaggi's small socialist group. VSP opposition to the Labor Party during 1911–12 ensured vigorous debates, at which Goldsmith excelled. Undoubtedly, Goldsmith would have been prominent among the local socialists who 'would be heard every Saturday night at the street corners, vigorously denouncing the Labor Party and all its ways'.[19] Goldsmith possessed the two attributes required to be an effective, influential rank-and-file member of the union. Acknowledged as an experienced, expert miner, he was also remembered by union veterans as a formidable fluent orator, who could direct the cruellest barbs at the established Council alliance between union leaders and local businessmen. It was Goldsmith, for example, who questioned the motives behind the number of local businessmen joining the Labor Party, memorably describing them as 'men who had swung on the whiskers of the workers'. In the lead-up to the 1912

---

19   D. Thorpe, *Criterion*, 30 March 1912.

Council elections he returned to this favourite theme. He questioned 'the right of the PLC or any other body to select men to represent the workers, for anybody could join the ranks of the PLC and pose as a labourite'. He went on:

> That game is played for all it is worth in Wonthaggi, where auctioneers, merchants, chemists and storekeepers and newsagents set themselves up as representatives of the workers. The union should see to it that the existing state of affairs is altered... [20]

Unfortunately for Goldsmith and his comrades, reconciliation between the VSP and the Labor Party in late 1912 cut the political ground from under them and the group collapsed, not to re-form until years of war had radically altered the political landscape.

The industrial and political debates of Wonthaggi's initial years reflected a rapidly maturing community. By 1914, visitors arriving in Wonthaggi were no longer travelling to the frontier. Most likely they had arrived by train at the town's new station, having slowly crawled past the State Mine's operations at the Station Area, past the headframes and winding gear, the powerhouse, and the stockpiles of coal ready for shipment to the city. The station and railway shunting yard defined the northern perimeter of the town's commercial centre, and Wonthaggi's hotels, grocers, drapers and newspaper offices were only a short walk away. Although boarding houses were not as important as five years previously, they remained an option for some visitors. Most, however, preferred to choose between the town's three hotels: Taberner's Wonthaggi Hotel, directly opposite the station; the Powlett, a wide-verandahed hotel one block to the south of the station; or the Caledonian, a spare red-brick building a further half block to the west. All were two-storied, befitting Wonthaggi's status as a growing mining community. Visitors with business with the Council would have proceeded further up Hunter Street, passing the cluster of public buildings, with the new post office, constructed in Queen Anne style, holding pride of place. Teachers coming to the primary school would have had to walk a full block to the west; those bound for the new technical school headed south instead. Travellers seeking commercial opportunities could walk the length of McBride Street. The tent town of 1910 had become a distant memory in a thriving community seeking to establish its place as the pre-eminent

---

20 For Goldsmith's 'whiskers' comment see *Criterion*, 30 March 1912; for his argument on working-class political representation see *Sentinel*, 28 June 1912.

business and social centre of south Gippsland. A significant number of these visitors sought out the State Mine itself, and not merely for business opportunities. Under the management of Broome, the State Coal Mine had established its own reputation for innovation and efficiency. Production increased significantly, reaching more than half a million tons in 1914–15, and productivity grew accordingly. Notwithstanding the concerns of local businessmen and the union, Melbourne's daily press and journals such as the *Mining and Engineering Review* could legitimately write of a go-ahead mine, one in which during 1913 more than half the coal won had been mechanically cut, a feat unmatched elsewhere in Australia.

Even so, public opinion of the mine's stability remained divided. Employment growth and increased production served to again raise the spectre of economic stagnation. With the Miners Union undertaking the nuts-and-bolts work of community building (the club, the dispensary, the hospital), the PLC-dominated council sought to adopt a broader, more strategic overview to address the economic future of the town. The realisation of Wonthaggi's true economic potential had proved a vexed issue from the earliest months of the State Mine's operations. Although the State Mine worked relatively small deposits (especially by NSW standards), many in Wonthaggi fervently believed in the existence of more-extensive reserves, and argued for further drilling programs to establish their existence. Similarly, many legitimately argued that the Railway Commissioners were focussed primarily on securing their own sources of supply, over the needs of both the town and the community. For the miners, economic growth meant security; while for others economic growth offered the chance to capitalise on commercial assets. A sense of urgency drove local imagination, and imagination was often given free rein:

> Steel rail-making works have been started at Newcastle, and it is quite on the cards that this, and other iron and steel works which are usually established in coal areas will be started in Wonthaggi…[21]

No amount of argument, geological or otherwise, could undermine such local belief. After 1912, agitation to commence public coal sales escalated, culminating in February 1913 in the establishment of the 'Sale of Coal to the Public League' under joint municipal and union patronage. The League's manifesto conveyed Wonthaggi's fear of economic stagnation and yearning for industrial growth:

---

21   Abrahamson and Son. 1915. North Park Estate, Wonthaggi (land sale brochure). Wonthaggi: Abrahamson and Son.

Figure 11. A Wonthaggi coalminers' float prepared for a Melbourne Eight Hours Day celebration, c. 1912–14
The messages are clear: the State Mine needs investment to fully support Victorian industry, while Wonthaggi miners deserve improved working conditions.
Wonthaggi Historical Society Collection

Figure 12. Funeral of Alf 'Nugget' Trewin, killed by a fall of stone while working in 3 shaft on 22 April 1914
The local press reported that 'the mournful procession was headed by the Wonthaggi Union Band playing the 'Dead March', followed by 500 miners marching …'
Wonthaggi Historical Society Collection

The citizens of Wonthaggi earnestly request your co-operation in what has become a public question of the most vital importance to every taxpayer and the future industrial prosperity of the State of Victoria. We refer to the Government supplying screened coal to the public from the State Mine. It is now a matter of common knowledge that in the Powlett area Victoria possesses a great national asset in the immense quantities of good coal in proved seams, that can be raised at small cost… The Railway Commissioners could establish depots in the large centres on the same principle as in New Zealand where the cost of coal to householders has been reduced from 30/- to 20/- [30 shillings to 20 shillings] per ton; and we feel assured that… it is possible to show the corresponding reduction here.[22]

The League attracted the support of the Council, the Miners Union, the PLC, local businessmen, and supporters of non-Labor parties. It represented the prewar high-water mark of community consensus and mobilisation. The League's campaign did not advocate independent union militancy but instead sought to persuade by the logic of its case, flawed though it might have been. However, neither public meetings nor delegations, eight-hour-day floats or the promise of cheaper coal in the metropolis could move the Railway Commissioners or, apparently, the German Kaiser. The outbreak of war in August 1914 inevitably led to global crisis and effectively relegated Wonthaggi's agitation for public coal sales to the sidelines for another decade.

---

22 *Criterion*, 13 February 1913.

Chapter 2

Fragile prosperity

Consolidation and expansion, 1914–1929

On Empire Day 1912 Victoria's Premier, Sir Alexander Peacock, reminded Wonthaggi's primary school children that even 'if they forgot everything else, he asked them to remember that their first duty was to learn to obey'.[1] It was inevitable that Wonthaggi would join in the wave of national fervour and Imperial patriotism that swept the continent. Participation in a global war initially emphasised the benefits of social consensus and co-operation. State Mine managers worked together with ardent unionists in the locally formed Citizens Defence League, while the Wonthaggi Union Band regularly played patriotic selections as a salute to local volunteers as large crowds cheered them on to the trains taking them to military camps on Melbourne's outskirts. Activities such as these reached fever pitch in the months following the Anzac landings in April 1915. Recruiting booths were overwhelmed; by February 1916 more than 700 men had enlisted from Wonthaggi alone. Patriotic funds raised thousands of pounds for comfort funds and other war-related purposes, with the union's regular levy on its members popularly acclaimed as the most important reason behind this local success; while the sacking of all German workers at the State Mine reflected both union and management insistence that such men should be removed and replaced with 'Australian-born'.

Despite Wonthaggi's embrace of war, the seeds of social discord were sown early. By mid 1915 recruiting campaigns had been supplemented by calls for military conscription. The rate of enlistment could not be sustained, and the Australian Imperial Force's insatiable demand for reinforcements had stripped Wonthaggi of the last easily accessible reserves. At a recruiting meeting in

---

1   *Criterion*, 26 May 1912.

January 1916, attended by Sir William Irvine, Wonthaggi's local federal Member of Parliament and an enthusiastic conscriptionist, the audience voted on a motion 'that in the opinion of this meeting the time has come to adopt compulsory military training'. In a foretaste of Wonthaggi's future political temper, the vote was lost. But at the same meeting, a miner in the act of enlisting also revealed a darker reality behind the town's recruiting success: 'it was better to go to war', he told the crowd, 'than work at the State Mine'.[2]

The issue of conscription divided the town, and just as surely split the Labor Party. The civic wing of the Labor Party ensured that the Borough Council adopted a pro-conscriptionist motion, with miner-turned-businessman Matthew McMahon an enthusiastic supporter. Significantly, only one councillor, unionist Archie Nelson, totally rejected the Council's position. The union endorsed his opposition, and sought to put a spoke in the wheels of the conscriptionists' case by a unanimous resolution passed at a general meeting in May 1916:

> This organisation is opposed to any form of conscription, and in our opinion, if any conscription is necessary, it must be preceded by conscription of wealth, such wealth to be conscripted until the war debt is paid off.[3]

In the wake of the union's organised opposition to conscription, the Victorian Socialist Party (VSP) was re-formed during a street-corner meeting in late autumn of 1916. Members devoted themselves to defeating conscription while prosecuting a wider political agenda, one suggested by VSP lectures on topics as various as 'The War and the Workers', 'Is Conscription Justified?' and 'Revolution'. The Miners Union remained at the core of opposition to conscription and the union's young tyros – the mine's wheelers – were now let off the leash. By October 1916, on the eve of the first conscription referendum, wheelers were systematically jeering and jostling recruiting sergeants in the streets and attempting to break up conscriptionist meetings. More than one such meeting broke up in dissent and brawling, with surging crowds giving the long-time militant Jack Goldsmith a wild ovation as he proclaimed that 'we, the citizens of Wonthaggi, are uncompromisingly opposed to the introduction of conscription in any shape or form in Australia'.[4] Wonthaggi decisively rejected conscription in October 1916 by a majority of more than 340 in a local poll of 2600, and repeated

---

2 For a report of this meeting see the *Sentinel*, 20 January 1916.
3 *Sentinel*, 12 May 1916.
4 *Sentinel*, 13 October 1916.

the result a year later at Hughes's second attempt to win public support for conscription. The response of one young miner to the events of the second conscription campaign reflects the changing political temper of the State Mine's workforce. He wrote of one meeting:

> There were two or three conscriptionists present – returned soldiers, and at question time they wanted to know how we were going to win the war without conscription. I felt like asking them, 'Who the bloody hell wants to win the war?' but that would have been indiscreet…[5]

As in so many other industrial communities, the conscription arguments of 1916 and 1917 broke established forms of social consensus and political partnership. In Wonthaggi, conscription destroyed the previously dominant Labor alliance. For many, the importance of supporting conscription, and by implication the war, superseded earlier Labor Party loyalties. By mid 1917 a steady stream of local party members, including many leading business identities previously prominent in Labor's cause, followed the example of prime minister Billy Hughes and moved across to join the new Nationalist Party. Not for another decade would Wonthaggi's Labor-led political alliance regain its prewar dominance.

The war had far-reaching effects on Australia's economy. Years of war proved a crucible for industrial and manufacturing growth. But for other industries, such as coal, the impact of war proved more difficult to interpret. Coal production declined on all major Australian fields, but the cause of this decline in Victoria was different to that in NSW. NSW mines depended on a flourishing export trade, one that collapsed following the outbreak of war, and both production and employment fell. Locally, State Mine managers told a different story, complaining that a lack of capital investment and the falling levels of skill among the mine's workforce were deadly blows to mining efficiency and the achievement of required production levels. Ironically, such deficiencies led to a period of consolidation and relative stability: more miners (experience notwithstanding) were now required to maintain production, while previously neglected developmental work, such as the McBride Tunnel, began to open access to additional coal reserves and in addition provide even more work for a growing workforce. The effect of enlistment on the mine's workforce had been reversed by 1917, and by the end of the war, total mine employment had grown by more than 250.

---

5   Bert Davies to J B Scott, 10 December 1917. Scott Collection, University of Melbourne Archives.

But such consolidation failed to promote industrial peace. Wider forces were at play, not the least the efforts of the new national coal miners' union to assert its authority and to respond to the increases in the cost of living stimulated by war. A national campaign demanding a uniform eight-hour working day (including travelling time underground), a general 20 per cent increase in pay rates, and additional overtime payments lasted for more than a year before an arbitrated settlement in November 1916 conceded the Miners Federation's demands.

While local Labor alliances dissolved, the Wonthaggi Branch of the Federation consolidated its national links. A number of historians have lamented that over the history of the State Mine, the Miners Union proved less effective due to a preoccupation with national events over local concerns. But any argument about the importance of the particularity of Wonthaggi is not valid. To argue such a case is to ignore the dynamics of both the industry and the union. From the earliest days of the union at Wonthaggi, national (and international) events informed local priorities, and vice versa. Recurring patterns of migration – of men and their families, of mining practices, of ideas and political opportunities – meant that however isolated Wonthaggi might have appeared, the truth was otherwise. The emergence of a national coalminers' union after 1914 only consolidated this tendency, and Wonthaggi miners remained committed members of the national union throughout the existence of the State Mine.

The national industrial campaign of 1915–16 confirmed this. Despite the fact that one central claim – an eight-hour day, including travelling time underground – was already established practice at the State Mine, Wonthaggi's miners participated enthusiastically in strike action to support the Federation's general demands. In Wonthaggi, as in other mining communities, the debates over conscription that racked the Labor Party took place against a backdrop of persistent and increasingly acrimonious industrial disputes. A three-week strike in March 1916 demonstrated the union's repudiation of the State Mine management's efforts to contain the dispute. By November the industrial situation had deteriorated further. Ignoring increasingly frantic calls by management to reduce industrial action in recognition of the generous conditions enjoyed by State Mine employees, the union instead withdrew funds committed to building a union hall to provide strike pay, signalling that Wonthaggi miners would remain on strike until the entire Federation returned to work. The prospect of a re-run of 1909 loomed, only this time without the possibility of any coal from the Powlett. The union deliberately raised the stakes, denouncing the mine management as 'a political benevolent society for the [state] Liberal government's protégés'.[6] Such insults aside, the success of this industrial campaign, in the face of both federal and state government opposition, cemented the authority of the Federation: in the words of Edgar Ross, it 'had at last come to stay'.

A powerful and permanent national union had an immediate effect in Wonthaggi. The language of industrial militancy became increasingly explicit, and was backed by action. In May 1918 the Branch endorsed the resolutions of the Third Australian Peace Conference, while simultaneously condemning Hughes's intention to attend the Imperial Peace Conference in London. In July a mass meeting joined a growing union chorus protesting at the judicial frame-up of 12 Industrial Workers of the World (IWW) members jailed in NSW for sedition and other crimes against the state, sentences seen by many in the labour movement as a vindictive response by the prime minister, Billy Hughes, to the defeat of his conscription proposals, and to growing union disenchantment with the war effort. Successful strike action, a single, powerful coalminers' union and the radicalising impact of war and conscription had combined to widen Wonthaggi's industrial horizon. New national campaigns demanding the abolition of the contract system of payment as well as the introduction of a six-hour working day reflected a new industrial reality. Some saw this militancy as the best means of defending hard-won conditions; for others it represented a form of 'industrial unionism' that would usher in the fundamental restructuring of Australian unions into the 'One Big Union' (OBU), a fundamental reorganisation of trade unions that collapsed individual unions into a single centralised union structure committed to the industrial reconstruction of society. In December 1919 Wonthaggi members voted 630 to 139 in favour of establishing the OBU, 678 to 74 in favour of the abolition of the contract system, and 699 to 57 in favour of a six-hour working shift. At the same time, the Victorian Branch began to recruit among the workers employed in the brown-coal open-cut mines of the Latrobe Valley. Although ultimately unsuccessful in the face of concerted opposition from the AWU, this, too, represented a further attempt to create a single union among all coalminers, whether working black or brown coal.[7]

These were years of profound change, and to many unionists the failure of general strike action in Sydney in 1917 had demonstrated the weaknesses of orthodox union organisation. As the pre-eminent labour historian Ian Turner writes:

---

6 *Sentinel*, 3 March 1916.
7 ACSEF Minutes, Powlett Branch/Victorian District, ACSEF, 19 December 1919; December 1923 (undated entries).

The obvious lesson of the [1917] strike defeat was the lack of any authoritative central organisation… and from the end of 1917 there was a new urgency and purposefulness in the plans for trade union reconstruction.[8]

Supporters of union reconstruction rallied around the idea of the OBU. Advocates of the OBU came to be called industrial unionists. This was a time of political ferment, and not everyone understood the OBU to mean the same thing. In Wonthaggi, veteran unionist and socialist Jack Goldsmith represented the union on the 12-man OBU Victorian Organising Committee, supported in his advocacy by the Federation's general secretary, A C Willis, who visited Wonthaggi in late 1918. Constrained by the need to meet often contradictory demands from competing unions and union officials, the OBU was to be called into existence by a series of conferences, votes and back-room compromises, driven by the many unions that sought to amend the concept of the OBU to protect their own identity. But not everyone believed that the OBU would be created by conference vote.

These were also years in which Wonthaggi's recently won reputation as a focus for the militant political and industrial activity that characterised the postwar labour movement acted as a magnet for many socialist and radical organisations, including many of the groups proselytising for the OBU. The small group of miners who met at the Federation's offices in Wonthaggi in May 1918 to consider 'the advisability of forming an organisation for the furthering of industrial unionism' certainly reflected this trend. Despite the open-ended nature of the question they considered, there was really only one answer for this particular group. The meeting had been called by a small syndicalist group known as the Workers International Industrial Union (WIIU). Descended from the IWW (or 'Wobblies') that formed in Chicago in 1905, the WIIU had broken with the Wobblies over the issue of supporting political action as well as industrial militancy. It was linked to the equally small Australian Socialist Party, and aimed at 'consolidating the workers into industrial unions on the basis of class struggle'. The 15 members recruited at a follow-up meeting on 1 June were joined in the following week by a further 17; by September the local branch claimed 150 members.[9] The WIIU shared common ground with the 'Official' OBU (the Workers Industrial Union of Australia, or WIUA), but remained sceptical

---

8   Ian Turner. 1965. *Industrial Labour and Politics*. Canberra: ANU Press: 182–183.
9   Minutes, Workers International Industrial Union Recruiting Local No 3 (WIIU), Statement of Objects, 6 May 1918; 19 May 1918.

of its reliance on incumbent union officials and its neglect of rank-and-file activism and on-the-job organisation to secure workers' control. The WIIU initially sought to improve its position through active educational classes among union members, but as the climate within Australian unionism swung decisively in favour of the Official OBU, it adopted a more antagonistic position. As a presumed agreement to recognise each other's members broke down, the WIIU implemented a policy of white-anting existing unions and rebuilding them in its own image. With limited resources, the WIIU concentrated its efforts in localities where it had some measure of support, however limited. Organisers were sent to step up recruitment among timber workers on the Western Australian goldfields, among AWU members working on the construction of the Eildon Weir, and among Wonthaggi miners.

Sensing an opportunity to establish a substantial permanent presence, the WIIU despatched an organiser, Jim Scott, to 'ginger up' the Wonthaggi WIIU Local (Branch). Brash, outspoken and more forceful in his opinions than most ('that forlorn coal camp' was his first impression of Wonthaggi), Scott didn't waste any time in tackling local Federation officials. Labelling rival plans for OBU reorganisation as a 'new abortion', Scott jettisoned the collaboration with the WIUA and the Federation's leaders that had marked the strategy of previous years in favour of an insurgent industrial campaign.

During September 1919 a pamphlet flooded the town:

WORKERS OF WONTHAGGI

ARE YOU CONTENTED?

To keep on working and only getting sufficient out

of it to gain the strength to go to work?

INDUSTRIALLY ORGANISED in the WIIU [the boss]

fears you because you are in a position to FIGHT and win![10]

Following Scott's strategy, the WIIU sought to win the miners' approval to restructure the Federation into a new industrial union. Despite a few prominent recruits, the majority of members remained indifferent and the WIIU feared, with good reason, that officials from the Federation and unions such as the Carpenters and the Engineers were now mobilising against their efforts. The WIIU played for time, believing that it lacked the strength for an open split, but time was not on its side. The Federation

---

10   WIIU. *Workers of Wonthaggi* (pamphlet), n.d. (August/September 1919).

brought on a motion to officially endorse the Official OBU: if passed, the WIIU in Wonthaggi would be effectively isolated. Led by Scott, the WIIU sought endorsement of its alternative strategy. The result proved closer than many in the Federation had anticipated. Only by 240 to 154 votes was the Official OBU endorsed. Outraged WIIU members proposed to refuse to pay union contributions or levies, but then, in Melbourne, the WIIU General Executive hesitated: it 'considered [the] matter of breakaway and... advised the local to stay [its] hands for the present and continue to recruit members'. Prevarication effectively ended any faint hope of success. The high-water mark of the WIIU had passed and local organisation collapsed: by mid 1920, even as national debate over the OBU peaked, the activities of the WIIU in Wonthaggi ceased.[11]

The letters of WIIU activists in Wonthaggi provide a rare window into the fears and hopes of working miners. Scott, no expert miner himself, foresaw an organisation that would become

> the organisation, not only to organise the town but to create that spirit of social camaraderie by concerts, card parties, outings, etc. and bringing together all the slaves, their wives and kiddies.[12]

Scott had sketched a vision of union organisation and community autonomy that would be realised, under far harsher economic circumstances, after 1934. In the meantime, the pressure of combining political activism with a miner's working routine had become only too evident. For some activists, Scott's forlorn mining camp had become a hell on earth. Bert Davies had acted on his own initiative in 1915 and cycled more than 800 miles from Broken Hill to Wonthaggi to pursue his own militant beliefs. But even for those as committed as Davies, the effort took its toll. On occasion imagination and fear outstripped political commitment, as when he wrote to Scott in late 1917:

> I couldn't face the pit today... I just simply couldn't face the track to work. The mouth of the pit became transformed into a hideous chasm and I shuddered at the thought... I sometimes wish I believed in God so that I could curse Him for driving me to this place... [13]

---

11   For the confrontation with the union Executive and the following collapse of the WIIU in Wonthaggi, see WIIU minutes; *OBU Herald*, July 1919; September 1919; Andrew Reeves. 1973. 'The rise and decline of industrial unionism: The Workers International Industrial Union in Australia'. B.A. Honours thesis. Melbourne: University of Melbourne.
12   J B Scott to 'Clark', 31 July 1919.
13   Scott Collection.

So why did the WIIU flourish, albeit briefly, in Wonthaggi? Why not in some other coal community, closer to Sydney, the centre of OBU organising? Part of the answer lies in the newness of the mine and its community, and the relatively recent embrace of radical politics by the workforce. Similarly, this was a workforce employed directly by the state, able to observe its centralised powers at close range. And there is always serendipity to consider: it was young men such as Davies, schooled in the rough and tumble of wartime industrial politics, who would form the core of the WIIU push in Wonthaggi. Ultimately the WIIU failed in Wonthaggi, as elsewhere. Members moved on, or returned to more orthodox day-to-day patterns of union activity in the town. But the WIIU did reflect the ferment of ideas and aspirations abroad within Australian unionism in the years following the First World War. By feeding a perception that industrial unionism represented a natural stage of development for unions, the actions of the WIIU ultimately strengthened the position of the Federation's established leadership.

It would be wrong, though, to view the union during these years simply as a hot-bed of agitation, with every second member a socialist agitator spruiking the new social order from a soap box outside Taberner's pub. The local leadership reflected a membership generally holding less ambitious political aims. John (Jack) McVicars represented an altogether different sort of politically active miner. Born at Ipswich in Queensland, the son of a Scottish miner, McVicars himself began work as a 14-year-old trapper at the Mount Kembla Colliery in Illawarra in about 1891. Two years later, with his father, he moved to Korumburra, where he worked at Coal Creek and became the secretary of the VCMA lodge at that mine. McVicars travelled briefly to Queensland in the wake of the disastrous 1903–04 strike but soon returned to the position of secretary of the breakaway South Gippsland Miners Association. Lured to Wonthaggi by the prospect of regular work, McVicars readily made his peace with the VCMA, being elected in 1913 to the joint office of Branch and Victorian District secretary, a position he retained until his retirement in 1946.[14] As with so many other Victorian miners of his generation, the experiences of 1903–04 heavily influenced his actions. He emphasised conciliation over confrontation. Although he did attend a conference of Victorian unions called in September 1918 to discuss the progress of the OBU in Australia, McVicars stood aloof from this wave of militancy, and it was due to his influence that during the 1920s the Victorian

---

14 See Andrew Reeves. 1986. 'John McVicars'. *Australian Dictionary of Biography*, vol. 10. Melbourne: Melbourne University Press: 366–367.

District of the Federation continued to support the principle of arbitration in contrast to the union's preferred policy of direct industrial action.

The first years of peace proved exasperating. Nothing seemed to go quite to plan: Wonthaggi was not a town fit for heroes. Union members had to choose between the faint promise of a new society offered by the OBU and the more tangible but lesser benefits of arbitration. Managers struggled to maintain production and to find the capital to continue necessary mine development. George Broome wrote that 'there are great difficulties in the way of developing the State Mine for a large and constant output, chief of which are the irregular and faulted nature of the deposit, which renders developmental work heavy and the cost of production high as compared to normal coalfields'.[15] But what was a normal coalfield? No-one had a ready answer to that question. Broome's point was merely academic. If Victoria were to have a black-coal industry, then it would be at Wonthaggi, faulted seams or not.

Mineworkers had their own views of the frustrations surrounding mine development and its inevitable impact on the town and, unsurprisingly, their views differed from Broome's. For them, these postwar years proved to be broken years, years of disappointment and anger marked by a revival of a mood of insecurity and impermanence. Miners believed that neither arbitration nor the state government addressed their legitimate industrial concerns or their essential security and that as a consequence they and their families were unnecessarily imperilled:

> They did not want to sell their homes and furniture, but they wanted to stop the boot being put in, as had been done [previously]… The position was unbearable, and it was a pity to see families stacking their all and leaving… All estate agency windows were full of places for sale. Men were wanting to get out of town.[16]

As late as 1925 the mine had to turn away hundreds seeking re-engagement, with concession fares provided for mineworkers wishing to leave the district.

As previously mentioned, the available positions at the State Mine were contested by and between returning volunteers, employees recruited during the war years to maintain the workforce, and growing numbers of British immigrants who brought with them their own mining skills, war experiences and union traditions. And the postwar economy could not break

---

15 George Broome. 1923. *Annual Report of the General Manager of the State Coal Mines, 1922–23*. Melbourne: Government Printer: 7.
16 *The Age*, 25 March 1923.

the repetitive cycles of dispute and interruption that marked the operations of the mine itself. The realities of coal mining deflected any sense that peace might actually mean a new start for all. Some detected a sense of sourness or a perception that the war years might have been wasted years, wasted in the sense that Wonthaggi and its staple industry could identify few benefits to offset the extraordinary human cost of war.

The reality that no-one had their hands on effective levers of control lay at the heart of this discontent. Although Wonthaggi miners looked to arbitration as the preferred way to settle disputes, these were by no means years of industrial peace. In 1922–23, 63 days were lost to strike action; 67 in 1923–24; 49 in 1924–25; and 53 the following year. But such apparent militancy disguises a deeper irony. If one accepts days lost to strike action as a simple definition of militancy, then Wonthaggi's miners were unwilling militants. They remained acutely aware of how dispensable the Railway Commissioners believed them to be. What drove strike action was not so much the desire to squeeze mine management but rather the need to reinforce repeated demands for a regulated industrial relations system at the State Mine, based on principles of conciliation. Many strikes, paradoxically, sought to force mine management into direct negotiations or, alternatively, to convince the state government to establish a State Mine Tribunal through which grievances could be aired and its National Award policed. A National Coal Tribunal had been established following passage of the Commonwealth Government's Industrial Peace Act in 1920. The National Tribunal addressed industry-wide issues, with day-to-day decisions that affected individual mines delegated to a series of state-based boards. While conscious of their place in a national industry and a national union, Wonthaggi miners also looked to a Victorian tribunal to address problems endemic to their relationship with the Railway Commissioners. By mid 1921 a union correspondent would complain that 'over two months ago, representatives were elected to the Wonthaggi tribunal… but no sitting had yet been held', while a further two years on, in March 1923, J J Long, an executive member of the Wonthaggi Branch, would be quoted in *The Age* defending strike action because 'mineworkers had good reasons weeks ago to strike but they had worked, hoping for a tribunal to settle disputes. The real cause of constant trouble lay in Melbourne. When justice was refused, the only thing was strike action, which should be obsolete'.[17]

Such ambivalence about the benefits of strike action as an alternative to negotiation or conciliation is illustrated by the strike of January to March 1923.

---

17   *The Age*, 25 March 1923.

As was so often the case in the coal industry, this dispute arose from differing interpretations of the current union award. Awards were intricate documents, particularly in regard to contract rates. Constantly fluctuating conditions at the coal face demanded an elaborate scale of rates and bonuses as a means of equalising earning opportunities. In November 1922 W J Dowling, as District president, complained to Broome of 'members being compelled to throw coal over a specific width', thereby substantially reducing the miners' production and, hence, their wages. The workplaces for individual mining teams underground were known as 'bords', a term brought to Australia from northern English mining fields in the nineteenth century. The narrower the width of the bord, the easier it was to cut and load coal (and, incidentally, the greater the number of miners employed). The width of the bord, and hence the effort required to 'throw' the coal into skips for transport to the surface, had a direct impact on miners' earnings. The union claimed 33 inches as the maximum width. Broome countered with a definition of 44 inches, thereby eliminating the need for extra payments to miners working in wider bords. Over the Christmas break, negotiations tempered the demands of both parties, but a final offer to the union of a 42-inch bord proved unacceptable, and the mine did not reopen after the New Year stand-down. The union had gone on strike. As a further tactic, the miners agreed not to return to work until a range of other grievances related to payment for development work had been settled as well. With appropriate publicity, the Railway Commissioners ordered the removal of essential rolling stock. The federal Tribunal finally acted in March, appointing an independent arbitrator to resolve the stalemate. Work resumed under pre-strike conditions. When handed down the following August, the arbitrator's determination amended the existing award by establishing a compromise of 40 inches as the regulation width for a working place, while also retaining existing overthrow payments.[18]

While not a particularly memorable strike, it remains typical in both cause and effect. Despite a wariness of strike action, miners would use it in order to defend wages and conditions. But such action also sought to bring about negotiation and consultation rather than to simply force a confrontation. In the case of this 1923 strike, once an independent arbitrator had been appointed, miners readily abandoned the additional demands they had used to escalate the strike in favour of a much higher

---

18 To follow the evolution of this strike, see ACSEF Minutes, November 1922 – March 1923.

priority – management acceptance of a structure for dispute resolution that included the union. In reporting back to a mass meeting relatively late in the strike, 'the delegates claimed that as far as [that] principle was concerned, it was a win and a complete breakdown of the stand taken by management and the Commissioners'.[19] Arbitration of disputes acknowledged equality between union claims and railway interests. For Wonthaggi miners in the early 1920s, the apparent neutrality of arbitration offered an attractive alternative to commissioners' intransigence. In line with this preference, after 1920 most militant members of both the District and Branch committees had been replaced with moderate Labor unionists. W J Dowling, who served a number of terms as District president, and Arthur Asquith, another executive member, together with secretary Jack McVicars, formed the triumvirate that dominated the Federation in Victoria until the Depression.

By 1925 the State Mine's operations were beginning to encircle Wonthaggi, giving visitors the impression of a mature colliery town. The mine's original, prewar pits – 9 Shaft, 10 Shaft and Station Area – had all been situated close to the mine's administrative block to the immediate west of the business district and the early, government-constructed cottages. War-time demand for coal had prompted the urgent development of the badly faulted yet substantial Eastern Area after 1916. With the adit opening up the eastern reserves running 4000 feet, working three benches (or levels) simultaneously provided a reliable supply of 500–700 tons per day. But the exhaustion of the prewar pits threatened to offset such gains. Production declines were halted, and then reversed, by the development after 1925 of the Dudley Area, tapping reserves to the north of Wonthaggi as far as the Melbourne road. The previously small settlement of South Dudley grew in to a suburb, accommodating miners involved in the Dudley development. Finally Wonthaggi had spread beyond its prewar limits to occupy most of the land optimistically surveyed and subdivided in 1911. While the bulk of the population still lived in an arc sweeping around the southern perimeter of the town's commercial district, many mineworkers and their families now lived in Wonthaggi's satellite suburbs – North Wonthaggi, South Dudley and Hicksborough. These grew appreciably after 1925 during years of substantial population growth. Formerly a small community with promise of potential, during the 1920s Wonthaggi developed into Victoria's fifth-largest provincial town.

---

19   ACESF Minutes, 15 February 1923.

Despite the effect of wartime recruitment, Wonthaggi's population had grown during the war years, reaching 4200 by 1918. In the postwar years, returning servicemen found themselves competing with the mine's own wartime recruits for available positions. Even so, an increase in developmental work and the greater number of miners required to work thinner, less-productive seams could possibly have absorbed most, if not all of them. However, such a relatively easy accommodation was confounded by the arrival of significant numbers of British immigrants, often experienced miners and ex-servicemen themselves, who also sought employment at the State Mine. These three sources of population pushed Wonthaggi's population to more than 7300 in 1930, or possibly as many as 10,000 if satellite suburbs and other nearby settlements are included.

The State Mine's expansion during the 1920s provides the only explanation for such growth. The mine's prewar average workforce of 1100 had by the 1920s grown to 1500. But railway coal consumption and natural growth, either singly or together, could not explain such an increase. Neither could the effect of thinner, poorer seams, however much of a reality they had become; nor even the peculiar rhythms of coal mining – that uncertain pattern of strikes, mining accidents, production problems and erratic economic demand. These were undoubtedly important, especially the impact of the fire of 10 August 1924 that gutted McBride Tunnel, killing four miners and spurring on another bout of developmental work, while the reintroduction of a three-shift system in unaffected pits maintained production. But this was transient, temporary work. In seeking to answer the conundrum of how to increase production to support both productivity and employment, attention turned, as it had before the war, to new markets for State Mine coal.

Opening up the Dudley Area raised, once more, the dilemma that bedevilled the State Mine and, by extension, Wonthaggi, for the life of the mine. The spectre of economic stagnation and unemployment could only be offset by increased investment, but capital investment resulted in the production of more coal than the Victorian Railways required. In turn, overproduction meant shorter hours or layoffs for the mine's workforce. Such contradictions threatened to undermine local confidence.

There was no easy solution to this dilemma. The preferred answer revisited solutions first proposed in 1912–13 that argued the need to systematically develop public markets for State Mine coal

> if the State Coal Mine, on which the town depends for its existence were developed on proper lines, the result would be a considerable

increase in population and a vast accession of wealth to the State. Under present conditions, the progress of Wonthaggi is being retarded because of the restriction of operations at the State Mine. [Instead], the Mine's Department should be empowered to work the mine to the limits of the requirements of the people of Victoria.[20]

Between 1923 and 1925 many in the town campaigned to remove control of the State Mine from the Victorian Railways. Although this campaign ultimately failed to deliver the mine from the Railway Commissioners, it succeeded, finally, in establishing a regime of public coal sales. Where threats had failed, increasing production, and the reluctance of the state government to curtail such increases, finally worked. In November 1925 public coal sales commenced and, much to the relief of the mine's manager, George Broome, quickly eliminated the margin between mine production and railway requirements:

> Since State Mine coal was made available for sale to the general public… there has been a gratifying and increasing demand, and it is becoming generally recognised that taking price and quality into consideration, State Mine coal is an efficient and economic fuel.[21]

After 1923, agitation for public coal sales provided the necessary impulse for the re-establishment of a local political alliance between Labor and business in Wonthaggi. But unlike prewar arrangements, a space had now been found for non-Labor politicians. The imperative to grow the State Mine and foster secondary, coal-dependent industries generally over-rode political affiliation. On the Borough Council William Easton, a pro-conscriptionist during the war, and an aggressive conservative in a Labor town, served as mayor between 1926 and 1928, with unanimous Labor and union support. Long gone were the conscription-fired days of 1917 when Labor councillors had publicly sworn never again to vote for any conservative standing for municipal office. Once a pariah, Easton had now become, in the words of two aspirant union politicians, 'the right man in the right place' who 'put the Council first, even if it was for the good of the borough against the policy of the party he was representing'.[22] Such a description could equally

---

20 *Express* (undated clipping, c. October–November 1923). Bill McKenzie scrapbook.
21 *Annual Report of the General Manager of the State Coal Mines, 1926–27.* 1927. Melbourne: Government Printer: 5.
22 The sentiments of endorsed Labor councillors J H Wishart and J Strong, *Sentinel*, 4 September 1925.

be applied to Labor councillors, men willing to accommodate Labor policy and principles to the perceived commercial priorities of Wonthaggi.

The election of W G 'Bill' McKenzie to the newly created state electorate of Wonthaggi in 1927 set the seal on this partnership. McKenzie had impeccable credentials: a life-long member of the Labor Party, he had carved out a successful career in the town as a draper after leaving employment at the State Mine over a dispute with management concerning the unionisation of mine clerks and other administrative staff. Achieving even greater prominence in the campaign for public coal sales, his defeat of both the District president and the secretary of the Federation in Labor preselection for the new seat starkly demonstrated the ascendancy of the re-established alliance between business and Labor. Few would have disagreed with the unbridled optimism of one local editorial:

> In the most emphatic manner possible, the electors of Wonthaggi have decided that Councillor W G McKenzie, the selected Labor candidate, is the most fitting person who could be chosen as their representative. Saturday last will long remain fresh in the memory of those whose interests and livelihood are centred in the district, for the reason that it was the day on which Labor came into its own... The fight is over, the gloves are off, and party differences will soon be forgotten when all work together in a common cause, namely, for the progress and development of the Wonthaggi electorate.[23]

During McKenzie's first term in parliament, Wonthaggi mineworkers enjoyed a rare sense of stability and security. The town's population steadily increased with the growth of the State Mine, and even if ancillary industries had not eventuated, locals still held hopes in that regard. In the eyes of local councillors and the local press, Wonthaggi's prosperity depended on industrial peace. But even the Victorian District's preference for arbitration could not eradicate industrial disruption and strike action. In taking their civic work beyond the council chamber, councillors themselves turned to time-honoured union forms of communication, particularly mass public meetings. Such meetings, capable of attracting up to a thousand people, became a regular feature of Wonthaggi's political life. Evoking Henry Lawson's description of the union 'roll up', the Wonthaggi Union Band often summoned citizens to these meetings. Under the aegis of the mayor, council or union, these meetings acted as public forums at which sectional

---

23   *Sentinel*, 15 April 1927.

differences could be laid aside and a sensible or reasoned approach agreed. So, for example, McKenzie could tell a crowd of 700 in January 1925 that discussed, among other issues, a protracted dispute with the Victorian Miners Tribunal that 'there was nothing burning or bitter in the dispute [which in fact there was] and that, after all what may be now a mountain would be found to be a mole-heap'.[24]

Such meetings also acted as a safety valve. During these years mineworkers' priorities generally coincided with those of local politicians, but not always. Relying by necessity on the state of the town's economy for their own prosperity, Wonthaggi's politically active businessmen at times found themselves more involved in political and industrial affairs than they may have anticipated. Some, like the Durham family – McBride Street grocers – had arrived when Wonthaggi had been barely a tent town. Others, such as Bill McKenzie and Paddy Webb, had successfully established businesses after years of residence in the town (and also notable union careers). Some were active in the Labor Party, others were not. Differing approaches to the State Mine proved a conundrum: as a group they leaned sometimes toward union claims, while at other times advocated a managerial position, often pleasing neither groups. For many businessmen, the hope that Wonthaggi might yet become another Newcastle still burned, and many believed that Wonthaggi sat on undiscovered coal reserves. At times, their interests paralleled those of the Federation, as was the case in 1927 when the Victorian District supported an abortive bid to have arbitration endorsed as official union policy. For the union's local allies, arbitration promised 'a continuity of work, happy and contented homes and a prosperous and progressive Wonthaggi'. Yet at other times this alliance could erupt into angry conflict, as was the case in 1927 when the majority of mineworkers elected to the local hospital's committee of management decided to deal solely with the union-aligned Co-operative Store for hospital supplies. While they had 'no quarrel' with the co-op, businessmen protested that: 'When it came to a principle every firm... should have the right to put in a tender for a public institution'. 'And exactly what principle would that be?' the union replied rather brusquely, reminding the spokesman for the business case, newly elected MLA Bill McKenzie, that 'one of the planks of the party to which he [McKenzie] was connected was the Co-operative Commonwealth'. [25] And there, on a point of unresolved disagreement, the matter uneasily rested.

---

24   *Sentinel*, 30 January 1925.
25   *Sentinel*, 2 September 1927.

By the time of his election to state parliament in 1927, McKenzie had become one of Wonthaggi's leading local identities, and over the next 20 years left an indelible mark on the town and its community. Although an active member of the local business community and long associated with the ALP, he often proved capable of standing above divisive community issues, acting instead as a bridge between the union and the community, as a political voice for the State Mine and as a tireless advocate for civic development and security. His electoral victory in 1927, repeated in 1930, typified a renewed sense of confidence and heightened morale in the town. With McKenzie's success, some claimed that 'Wonthaggi's time had come', for the prospect of Labor in power had come to be understood as synonymous with economic progress. As 'Wonthaggi's mentor', McKenzie's success exemplified these all-too-brief years of prosperity that preceded the onset of depression. At the mine, average earnings after 1926 remained stable, while aggregate wages increased steadily, due to the twin effects of increased production and fewer days lost to strike action. Drilling and other exploratory work had been pushed ahead in the mine's Western Area as well as in the Kirrak Basin to the east along the Inverloch road, while McKenzie continued to promote public coal sales and the development of light industry in the town as the basis for continuing prosperity. It would be too easy to suggest that during these pre-depression years Wonthaggi lived cocoon-like, preserved by local conditions from wider economic influences. Despite a sense of increased security, many in the town, McKenzie included, remained acutely aware of pressures building against Wonthaggi's fragile prosperity. McKenzie was no simple booster for his community. He proved himself, time and again, to be an astute politician, with his eye fixed on tangible results. McKenzie believed Wonthaggi to be in a race against time: a small coal town in a 'world getting beyond coal', with fine seams that might never be mined unless developed immediately. Markets for State Mine coal had to be expanded and diversified to withstand periodic slumps or the whims of the Railway Commissioners.[26]

One early sign of trouble ahead proved to be Broome's decision in late 1927 to reduce the mine's operations by one day a week to reduce coal stockpiles. Immediately, local opinion dismissed his decision as of little more than peripheral concern. By arguing that short-time stemmed from overproduction rather than strike action, the Federation displayed that its main interest

---

26 So too did conservative Councillor Easton. For examples of their arguments see *Sentinel*, 18 February 1927.

lay in demonstrating its innocence. Others proved equally unconcerned, claiming that with increasing coal production, a temporary glut might at times be inevitable. Informed local opinion held that overproduction would not be a permanent problem if both government and industry accepted that the State Mine represented a valuable state asset that retained money in the Victorian economy if manufacturers acted accordingly. 'If manufacturers would give greater support to the mine and purchase less coal from other states', McKenzie vehemently argued, 'there would be no need to close down because of overproduction'.[27]

If local miners and their families required further reassurance, the growing success of their own commercial activities provided comforting evidence. The most spectacular success had been achieved by the Wonthaggi Co-operative Society Store. Established in 1912, following complaints about the local cost of domestic commodities, the co-op drew the bulk of its shareholders from the ranks of miners and other unionists. Through its members, the Federation's Victorian Branch acted as a virtual guarantor of the co-op. Shares cost one pound each, and shareholders were required to hold at least ten shares, up to a maximum of one hundred. The co-op's first years proved erratic and disappointing, marked by low rebates and lower community interest. Things changed in 1922 with the appointment of John Short, an experienced Scottish co-operator. As manager, Short restructured the co-op, adding a bakery and butcher shop while also establishing contract arrangements with other local businesses that gave co-op members access to purchases as diverse as hairdressing and motor repairs, photography and funerals. Rebates rose dramatically, from an average of 1.25 per cent before 1922, to more than 16 per cent by 1928. The annual October Dividend Sale (timed to coincide with a distribution of rebates to members) became one of Wonthaggi's premier business events, and the biannual distribution of rebates served to stimulate the local economy. The Workingmen's Club had similarly prospered, and by the early 1930s about 500 members enjoyed not only its well-appointed bar and dining facilities but also its weekly dances and raffles.[28]

The functions of other co-operative ventures were no less important, even if less visible. Collectively, the Dental Clinic, Union Dispensary and Friendly Society provided a safety net for miners and their families

---

27   *The Age*, 22 September 1927.
28   Wonthaggi Co-operative Society. 1922–38. *Half-Yearly Reports and Balance Sheets*. Wonthaggi: Wonthaggi Co-operative Society; 1934. *Powlett Express Anniversary Souvenir*. Wonthaggi: *Powlett Express*.

unmatched in many other communities. But perhaps the greatest evidence of mineworkers' confidence in their future could be seen in the imposing Union Hall and Theatre, completed in 1924. Facing the Workingmen's Club, the hall dominated the eastern approach to Graham Street, Wonthaggi's major commercial thoroughfare. Capable of seating an audience of 1000 and decorated in a contemporary, lavish theatre style, the building was sumptuously furnished. As anticipated, the hall provided a focus for the union's industrial and political activities but necessarily served a wider community, competing with the Crystal Palace and the Soldiers' Theatre for local patronage. The facade of the Union Theatre promised 'pictures and dancing', and from the time of the first 'grand masked ball' in May 1925, celebrating three queens: the Queen of Scots, the Queen of Miners and the Queen of Wonthaggi (with the music provided by Mrs Connelly's Orchestra), the theatre fulfilled that promise. Movies ('the flicks') began a fortnight later, on Pay Saturday, 5 June. In their invaluable account, *It's On at The Union*, Lyn and Joe Chambers tell of Saturday and Sunday nights being the most popular for films, with the Wonthaggi Union Band often playing outside the theatre prior to Saturday-night screenings. Saturday afternoon matinees were undoubtedly the most unruly sessions:

> there was strong audience participation… when groans, cheers, catcalls, boos, hisses and at times the drumming of hobnail boots on the bare jarrah floor, would drown out the music. Even from adult audiences, music of a martial nature would always get a response.[29]

Friday night, especially Pay Friday, was a time for shopping (with shops open until 9 pm), casual conversation, gossip, and the entertainment offered by street preachers, political meetings and occasional brawls, with a keen eye kept by most on who had arrived in town that night on the evening Melbourne train. The Union Theatre hosted boxing on Friday nights before big crowds, with the prizes on offer attracting Melbourne professionals to take on the willing locals. As well as dances, movies and professional boxing, the theatre hosted political rallies, children's fancy-dress balls, theatre, fetes, bazaars and jumble sales. Within months of its opening, the theatre had successfully become a focus for community entertainment and political activity.[30] As a coastal town, the beach also occupied an important place in Wonthaggi's leisure activities. Parties, picnics and other social

---

29   Joe and Lyn Chambers. 1982. *It's On at The Union*. Wonthaggi: Wonthaggi Historical Society: 14.
30   For the importance of the Union Theatre see especially Chambers, *It's On at The Union*.

events were held on beaches along the Bass coast from Inverloch to San Remo throughout summer, and at some of these sites, such as Inverloch and Harmers Haven, beach shelters built over the years by mining families grew into comfortable holiday shacks.

The State Mine's expansion during the 1920s had enabled many British immigrant miners to go straight 'on to coal' and to accrue both seniority and authority over other mineworkers. This advantage was increased by the effect of the single cavilling system on the structure of the mine's workforce. The parties of four entering this ballot arranged and rearranged themselves, their composition determined by personal preference and agreement. Working in two shifts of two miners, each party equally divided the payments won for the entire output of the party. Although as a system of random work distribution cavilling served to prevent victimisation and/or favouritism, it did demand a great deal of trust and co-operation between small groups of miners, acting as an extension of their working and personal relations. British miners instinctively sought co-operation from among their own number. Cavil teams made up of fathers and sons, of brothers, or of miners from the same British mining community became increasingly common after 1922 as the number of immigrant British miners employed at the State Mine grew. The four Brydon brothers, for example, originally from West Fife, formed one of the most productive mining teams during these years, while other personal friendships that would later assume direct political significance were also established, or matured, in the State Mine's pits.

In the eyes of many miners, Wonthaggi's cavilling system helped to underpin much of the mineworkers' militancy. In contrast to the 'butty' system of subcontracting, or even the system of cavilling small, individual pits that so many Scottish miners well knew, the State Mine's single cavilling system covered the entire mine, from the Western Area to the Kirrak Basin, in a single ballot. No miner, however stable his party, could be confined to meeting and working with a limited number of other miners in his time underground. To the contrary, in the course of three or four years work underground a miner would, in all probability, work in most of the mine's productive and developmental pits, and would work directly with hundreds of other miners. Furthermore, the importance of crib time (meal breaks) underground and the deserved reputation of many British miners as crib-time orators enhanced the significance of the single cavil in assimilating the British miners and welding the State Mine's workforce into a cohesive industrial force.

Figure 13. Aerial view of Wonthaggi looking south-west, c. 1935–36
Graham Street runs diagonally from lower left to upper right, past the Union Theatre (sixth building from the bottom on the right-hand side of Graham Street) and the town's business centre, through the protective screen of pine trees shielding the town, to the State Mine administrative centre and the railway yards in the Station Area.
Wonthaggi Historical Society Collection

Figure 14. Saturday night outside the 'Union', early 1930s
Wonthaggi Historical Society Collection

Figure 15. Looking east along Graham Street a decade after the establishment of Wonthaggi, c. 1918
The stark-fronted hotel on the right is the Caledonian Hotel. The Powlett Hotel can be seen in the distance.
Wonthaggi Historical Society Collection

Figure 16. 'It's On at the Union': a boxing grudge match staged at the Union Theatre in 1929
Note the event is held on Pay Friday, ensuring a full house.
Wonthaggi historical Society Collection

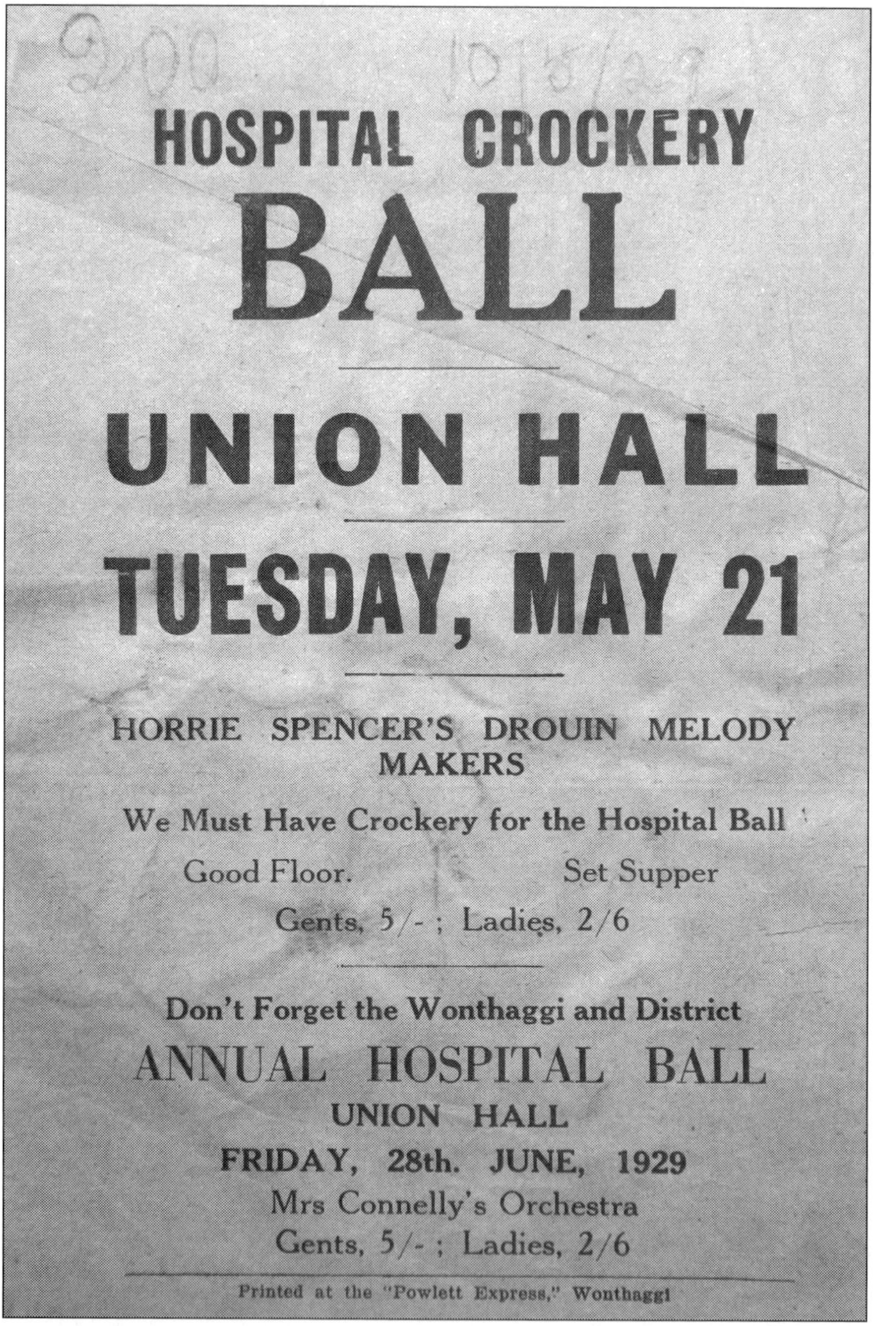

Figure 17. An advertisement for a charity ball at the Union Theatre in support of the local hospital, May 1929
Wonthaggi Historical Society Collection

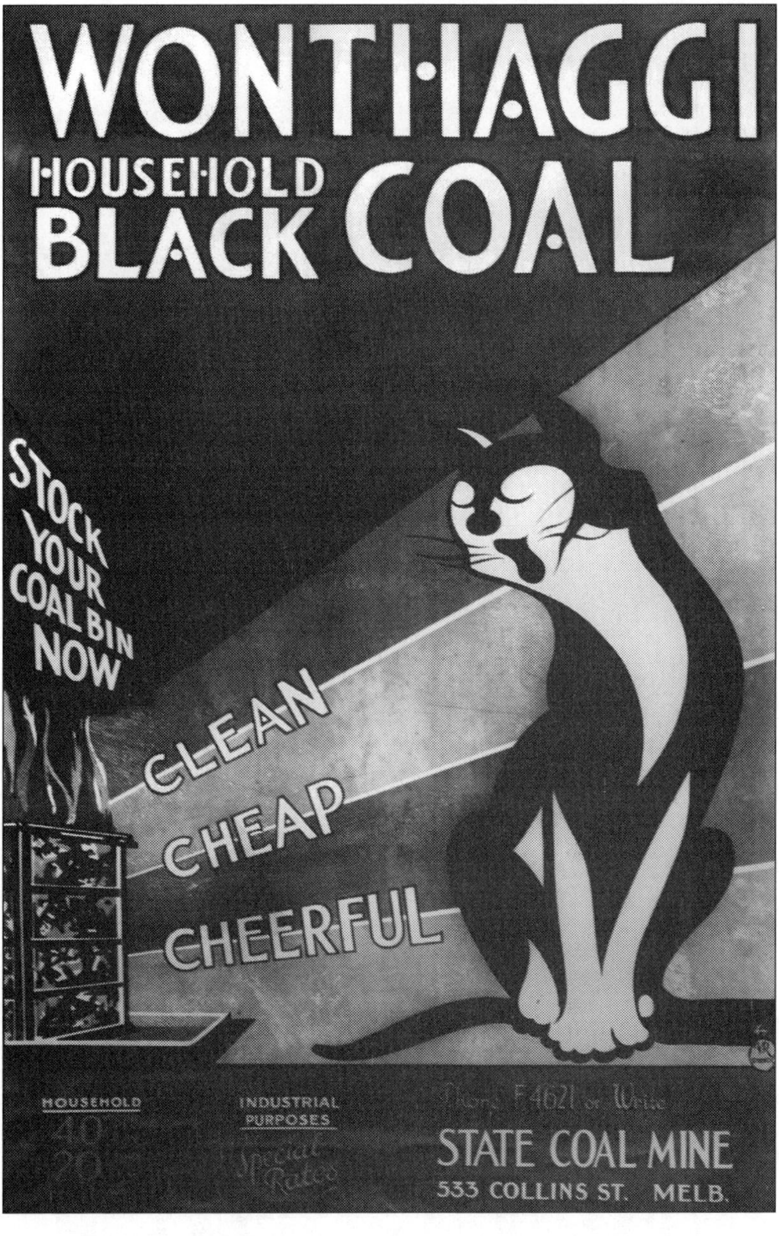

Figure 18. 'Wonthaggi Household Black Coal: Clean, Cheap, Cheerful'
A State Coal Mine advertisement promoting domestic coal sales in the Melbourne market as a result of community pressure for the mine to develop new markets, c. 1925–27.
Wonthaggi Historical Society Collection

Figure 19. The Wonthaggi Union Band parades in Wonthaggi, c. 1920s
Private collection

Figures 20 and 21. Wonthaggi cyclists
Cycling became one of Wonthaggi's boom sports during the Depression. The Wonthaggi–Melbourne road race became a 1930s cycling classic. These phtographs show riders marshalling in front of Taberner's Hotel.
Rankine Collection, University of Melbourne Archives

Social relations reflected such work relations. In particular, interdependence among migrant families was strengthened through intermarriage and other personal ties stretching back to Scotland, Wales or the north of England. For example, the Chamberses, Fosters, Hamiltons and Stirtons were as closely linked by empathy as by marriage. Blacklisted by Lanarkshire and West Lothian mineowners in the wake of the 1921 coal strike, Jim Chambers and Joe Foster left Broxburn, West Lothian, for Australia, nominated for emigration by a relative of Jim Chambers who was farming land in north-eastern Victoria. The two were brothers-in-law, Jim Chambers having married a sister of Joe Foster. Agricultural work proved unattractive, and, through correspondence with another Broxburn migrant, Chambers and Foster obtained work at the State Mine in 1923. During the course of the next three years, Jim Chambers would be joined by other members of his family, including his parents and two children who had continued their education in Glasgow. One of his daughters married Bob 'Hammie' Hamilton, a West Fifer who worked in an all-Scottish cavilling party with Bill Stirton, yet another miner from Fife. Jim Chambers's son Bill, a wheeler at the mine, intended to marry Nancy Grieve, daughter of another Scottish mining family who had emigrated to Wonthaggi in 1926, but before their marriage Bill Chambers died from the effect of injuries sustained in a mining accident. Nancy, already 'family' to the Chambers, later married Bill Stirton and the Stirtons, while not directly related to the Chambers, have been kin ever since. This is not to suggest that these migrants formed a closed community. The opposite is true. But they succeeded in creating social bonds among migrant families independent of the male work of underground mining. These were communal links on which much of Wonthaggi's militant union leadership would later be established.

During these years migrant families arriving in Wonthaggi had relatively little difficulty assimilating into the political and industrial loyalties of the town's established workforce, while consciously seeking to retain their own distinct identities and communities of interest, broadening and enriching local patterns of social life in the process. It is true, though, that in 1927 a short and bitter debate, possibly driven by tightening employment conditions and a local housing shortage and largely conducted through the Victorian Parliament, raised the issue of preferential employment for such migrants at the expense of Australians. But by that time such concerns could be managed within the local community. Such migrants had begun to imprint something of their culture and activities on the character of Wonthaggi. The town, for example, now boasted its own soccer league. Only on Australian coalfields, it seems, could soccer rival rugby or Australian Rules Football at that time. In Wonthaggi, too, it remained a sport with appeal to specific groups, as the team names 'Wonthaggi Thistles' and the 'Caledonians' suggest.

Idris Williams, prevented by a wartime injury from playing football (of any code), acted as secretary of the East Wonthaggi Australian Rules Football Club during its years of premiership success in the late 1920s, combining this with his position as band leader of the Wonthaggi Brass Band. The Wonthaggi Social Club reflected the increasing influence of this migrant group on community entertainment. The club survived and flourished due to the influence of British migrants, organising community recreation as well as the community singing that became a hallmark of Wonthaggi life in the 1930s. Dances were held up to three times a week and card games played for small wagers also proved popular. By 1933 its predominantly British-born committee included communists and ALP members as well as those with no political affiliation, housewives, union committeemen such as Bill Stirton, and an instinctive historian, Bill Rankine. During the 40 years following his arrival from West Fife in 1922, Rankine collected union and political journals, pit-papers, pamphlets and handbills, social and community material and co-op records that provide a unique insight into Wonthaggi life during these years, and on which this book draws heavily. The social club's executive positions were deliberately shared between men and women. While the fact that 'the ladies of the club… from time to time decorated the hall… and continue[d] to do great work… for the pleasure of the members' reflected established patterns of social relations, the club offered women opportunities to organise and engage in recreation that previously all too often had remained tied to the male worlds of the Workingmen's Club and the local pubs. Such experience was put to good use during the 1934 five-month strike when women occupied key organisational tasks with the Entertainment and Relief committees, as well as joining mineworkers in the propaganda teams that took the Wonthaggi miners' case to the rest of Australia.[31]

During the relatively prosperous 1920s another immigrant community had also gained a foothold. By the onset of depression an identifiable Italian community existed, concentrated at the western end of Hagelthorn Street. A significant number found employment at the State Mine, to the extent

---

31   For the growth of this migrant community in Wonthaggi, see Andrew Reeves. 1986. '"Damned Scotsmen": British migrants and the Australian Coal Industry, 1919–49'. *Common Cause: Essays in Australian and New Zealand Labour History*. Sydney: Allen and Unwin: 95–99.

that the Victorian District had a number of rules and regulations printed in Italian at this time.

Six hundred miles to the north, the onset of depression left the Federation's powerful Northern District locked out of the pits, vainly attempting to stem a concerted drive by employers to reduce both coal prices and miners' wages. While the inevitable national coal shortage sheltered Wonthaggi well into 1930 from the industrial storms that blew across Australia's fragile economy, the circumstances of Wonthaggi's final year of pre-depression prosperity are paradoxical. At a time when the Northern District of the Federation was facing political annihilation, Wonthaggi unionists could gain some comfort from being told that 'at no period in the history of the field has there been greater prosperity and contentment than exists today'. Even so, locals watched the bitter NSW dispute with trepidation, reacting angrily to a suggestion that the Commonwealth Government should subsidise a reduction in the price of NSW coal as part of a peace settlement – 'an insidious attempt… to strangle the coal industry in every state except NSW', McKenzie claimed. Although contributing a compulsory 12.5 per cent levy in support of Northern District mineworkers, the Victorian Branch took no active role in the strike. During the 16 months of the lockout, Wonthaggi miners stopped for three days, and then only in the angry aftermath of the Rothbury shootings. A suggestion floated by the state government to import coal from Britain only served to remind Wonthaggi miners that, although local production remained high, a decline in developmental work at the State Mine had already laid off a number of mineworkers. As one town councillor observed: 'there were men walking the streets of Wonthaggi looking for work, and if the Government did its duty… it would spend more in the large black coal deposits of Wonthaggi'.[32] Under local pressure the Railway Commissioners did increase daily production to 2600 tons, the mine's absolute maximum. Even so, railway requirements far outstripped the mine's capacity. In August 1929 the commissioners reversed their policy on the public sale of coal, immediately cutting such sales by 75 per cent. That the commissioners also took the opportunity to raise their coal prices, in defiance of national trends, only served to rub salt into raw, local sensitivities. McKenzie countered by arguing that such a regressive decision threatened miners' jobs, and foreshadowed a three- or four-day working week at the mine, but the commissioners remained unmoved. These manoeuvrings took place

---

32   Councillor Strong, *The Age*, 3 July 1929.

against the backdrop of the Northern NSW lockout, a dispute that once over would mean that railway demands on the State Mine would again be reduced. Few could even hazard a guess as to how long it would take to win back a public market.

At a glance, the State Mine had never appeared as prosperous and secure as it did during the first six months of 1930. Production exceeded previous figures and revenue had never been higher. Employment had peaked at over 1800 men. But such reasons for confidence were temporary. The change in the tone of the local paper – previously satisfied, now increasingly cautious – best captures the mood of trepidation that now captured Wonthaggi:

> The town has experienced three years of industrial peace, peace which has brought contentment to all parties concerned. Let us hope that it will remain so for many years to come.[33]

---

33   *Sentinel*, 19 July 1929.

Chapter 3

Up from the underworld, 1929–1934

Wonthaggi's prosperity could not last. By June 1930 union resistance to wage cuts on the Northern NSW fields had been broken, and production resumed. Inevitably, the State Mine took a hit. From a high of nearly 59,000 tons in February, by November monthly production had slumped by more than 30 per cent to barely 40,000 tons. But even such a severe slump could not convince the Railway Commissioners to again release coal for public sale. Without a prolonged increase in government and public demand for State Mine coal, miners and management found themselves once again caught in a closed cycle of declining production and intermittent employment from which there could be no escape. As veteran miner Bill Stirton recalled many years later, 'ultimately the demand for Wonthaggi coal declined and the miners, for a period, worked intermittently, sometimes four days a week, sometimes three'.[1] Mine revenue also fell steeply. The commissioners had previously tied the price for State Mine coal to prices set for Maitland (NSW) coal, and so revenues from the State Mine during these years fell in proportion to declining NSW values. By 1932 revenue from the mine had fallen to 1918 levels, despite producing an extra 180,000 tons.

Having rejected the possibility of increased production and increased public coal sales as solutions to the mine's revenue problems, the Railway Commissioners now moved rapidly, making an application in June to the Commonwealth Court of Arbitration for wage reductions at the State Mine. Judge Beeby found the commissioners' arguments of a need to reduce production costs to continue to compete with Northern proprietors compelling, and on 1 August delivered an interim award that cut wage rates by 12.5 per cent, with a smaller, 6 pence per day reduction for day labour. Although rejecting Federation arguments against such cuts, he nevertheless suspended their implementation for three months to allow a possible local

---

1   Bill Rankin. c. 1974. 'Depression in Victoria', mss notes provided to the author.

agreement to be struck. Such an agreement never eventuated. The Federation had turned to the state government, led since 1929 by the Labor Party, to conduct an inquiry into State Mine finances as a strategy for heading-off the wage cuts. The Premier, E J Hogan, demurred, instead telling State Mine employees that he 'felt obliged to subscribe loyally to the decision which [the Court] gave'. Hogan's failure to support miners and their union was neither forgiven nor forgotten in Wonthaggi.[2]

The effects of wage cuts belatedly imposed in November were soon felt in the pits and throughout the wider community. With wages reduced by as much as 40 per cent due to the compounding effects of intermittent work and wage reductions, a self-imposed speed-up developed as anxious miners used every hour 'on coal' to earn what they could from reduced contract rates. During 1929–30 the mine's fortnightly wages bill exceeded A£20,000. By June this had fallen to A£12,500, and by the end of 1930 barely averaged A£10,000, with the local economy an early casualty. Average weekly earnings for a skilled miner had fallen from A£7 per week to somewhere between A£2 and A£3 per week in a little over a year. 'They cannot carry on at that rate of pay', warned Bill McKenzie, 'and I am afraid that trouble will develop at the Mine'.[3]

During these years of deepening depression, when the appeals of hard-pressed industries all but drowned out the cries of threatened unionists, McKenzie's was the sole public voice raised in defence of Wonthaggi miners. Local Federation leaders – W J Dowling, union official and ALP Branch president; McVicars, well established as Federation District secretary; and, more recently, Bob Russell, a Scottish migrant of the early 1920s now elected District president – were all staunch Labor members, and represented the base of McKenzie's political support. But faced with economic pressures unknown for almost two generations, these union leaders increasingly looked like rabbits caught in a spotlight. Although his own position was tempered by 'economic necessity', McKenzie fought to preserve both the union leadership and the miners' working conditions.

McKenzie's best efforts ultimately proved ineffectual: nationally the Federation was in danger of collapse, with the federal Council estimating that by June 1931, 40 per cent of members were unemployed and the balance were mostly working only intermittently. The Victorian District remained solvent, and had recently supported miners on strike at a private mine in

---

2   *VPD*, 1930 Session, 6 November 1930: 3602.
3   *Sentinel*, 14 November 1930.

Kilcunda, 20 kilometres west along the coast, but only the late onset of depression had allowed such generosity.

Increasingly the local opinion was that Labor had failed the town. Unsuccessful attempts had been made in 1929 to convince McKenzie to run for the federal electorate of Flinders, which, to the astonishment of most, was ultimately won by the secretary of the Melbourne Trades Hall Council, E J 'Ted' Holloway, defeating the incumbent prime minister Stanley Melbourne Bruce. Not until John Howard's loss of Bennelong at the 2007 election did any electorate repudiate a sitting prime minister in this way. But neither a federal Labor government nor its state Labor counterpart could come to Wonthaggi's rescue after 1930. The optimistic belief of 1930 that 'Wonthaggi's future can be depended upon to be safeguarded while the [state] Hogan Government is in power' had deteriorated within 18 months into a sullen acquiescence, sharpened by bitter disillusionment and an inability to understand Labor's failure to reverse, or even halt, Wonthaggi's dramatic economic decline. Labor's loss of prestige paralleled the decline in miners' wages. When in February 1931 the Minister for Railways, John Cain, agreed to increased coal purchases from NSW despite falling production at the State Mine, many in the town were finally, bitterly, convinced. As the second local paper the *Powlett Express* put it, 'with the exception of a few of the [government] members, the workers will have to fight their own fight'.[4] The federal Labor government's decision to implement the Plan (and its additional wage cuts) only exacerbated a political crisis already out of control. The Federation, in common with other unions, denounced the Premier's Plan. McKenzie, as a Labor parliamentarian, equally vigorously supported it, notwithstanding its effect on wages. Scullin's federal Labor Ministry fell in late 1931; Hogan's Victorian Labor government followed in May 1932. Never again would the Labor Party enjoy such unchallenged support from Wonthaggi's miners as it had during the 1920s.

But where to turn? And on whom to rely? Factories were working half-time; hospitals were buying NSW coal; the Railway Commissioners were running fewer trains. McKenzie lamented Wonthaggi's decline: 'I have seen sights in my own town which I shall never forget, and my town was comparatively prosperous until the last twelve months'.[5] The State Mine no longer returned a profit, so its very existence would now be weighed against its continuing benefit as a local coal reserve and a defence against

---

4 *Powlett Express*, 13 February 1931.
5 *VPD*, Second Session, 16 June 1932: 97.

price exploitation by NSW producers. With coal at historically low prices, the latter argument seemed threadbare, all the more so as the Railway Commissioners now sought even higher prices for local coal in an attempt to restore profitability. In the words of the Borough Council, 'Wonthaggi wanted an assurance that there is a future for black coal'.

The effects of declining production and a reduced payroll at the State Mine inevitably flowed through to the local economy. Although muted by the absence of any major job losses at the mine, by July 1931 McKenzie estimated local unemployment at more than 200, with that number mostly made up by young single men. The despatch of clothing and food for Melbourne's unemployed had become a thing of the past, and working women now found themselves under pressure to relinquish their jobs in favour of unemployed men. The winter of 1931 turned out to be one of the worst in living memory: bitterly cold and continually wet. While deeply distressed by the effects of poverty on the local unemployed, McKenzie was increasingly concerned at the political dimension of unemployment:

> In a climate like that of Gippsland it is impossible for a man to walk down the street without getting his feet wet. The boots that the young men wear are broken and down at heel. Many of them have no socks, and their clothes are in tatters. It is bitterly cold at night. I want [you] to realise the danger to the State if this condition of affairs is allowed to continue. These young men are the most fertile soil in the world for the seeds of Communism... [6]

With the bulk of unemployment relief funding concentrated on Melbourne, local organisations sprang up to support local efforts. After mid 1931, an Unemployed Workers Union, supported by the Council and local ministers as well as the Federation, argued for improved financial support while also negotiating free entry to football matches and to films at the Union Theatre. Similarly, the town's co-operative businesses extended additional support. As an example, the Union Dispensary continued to provide benefits free of subscription to all unemployed members.

As a general recession settled on Wonthaggi, fewer and fewer industrial and political reports appeared in the local press. Depression numbed the town. Popular sports such as cricket, football and soccer filled columns of newsprint, while the depression's boom sports – boxing and cycling – attracted attention from miners and the unemployed with time on their

---

6   *Powlett Express*, 10 July 1931.

hands. So, too, did the re-opening in September 1933 of the Powlett Hotel, Wonthaggi's oldest hotel, under the proprietorship of the Richmond Brewery, with its slogan 'the Worker's Friend'. A crowd of 700 heard McKenzie talk of his confidence in the future of the town. While this no doubt gladdened the unemployed present, they were probably more gratified by the free meals the hotel provided for three days that fed Wonthaggi's destitute for a week.[7]

By mid 1932 Wonthaggi had endured two years of depression. Growing unemployment, short-time at the mine, reduced wages: all these issues served to sharpen one question – would the State Mine be closed? And if the mine closed, what then of Wonthaggi's future? Faced with unrelieved disaster, the Federation had concentrated on preserving jobs. Even jobs shorn by short-time and wage reductions were preferable to unemployment. In addition to the obvious economic arguments for preserving jobs, union members also saw the issue of employment in a moral light: hadn't they continued to work throughout the Northern District lockout when the need for coal was extreme? Didn't that count for anything? In their eyes, the decision by George Broome and his deputy mine manager, John McLeish, to retrench more than 400 mineworkers in July 1932 represented a gross betrayal. They were not chattels simply to be disposed of at the whim of mine managers.

But times had changed. There were now those politicians who questioned the very existence of the State Mine, particularly as it no longer returned a profit to the state. Furthermore, the guard at the State Mine had now changed, or, at least, change was well underway. By 1932 Broome had been general manager of the State Mine for nearly a quarter of a century. Like his counterpart for so much of that time, Federation Secretary Jack McVicars, he recognised and readily accommodated the rough-and-tumble nature of coal-mining politics. Broome also believed in arbitrated settlements as his guiding principle for industrial relations with the Federation. Like the union, he accepted the legitimacy of conciliation, even if his perspective was not that of the union. McLeish held different views, and would soon assume Broome's position as general manager. Broome had created the mine, whereas McLeish believed was to be to run a commercial enterprise. Broome often saw issues through local eyes, whereas McLeish applied a wider, more political perspective. As one veteran miner related many years later, Broome had a soft heart, while McLeish had a hard head. By mid 1932, McLeish successfully drove rationalisation at the State Mine. He prosecuted a two-fold strategy: firstly, cut wages: 'every possible economy

---

7   See the account in the *Powlett Express*, 9 September 1932.

has been affected in operating costs… The only way to … relieve the burden of [the mine's losses] is by a substantial reduction in wages'.[8] Secondly, McLeish favoured a further retrenchment of 330 employees. To head off any industrial response he suggested that those continuing in employment work an 11-shift fortnight, effectively guaranteeing adequate wages. In pursuit of these aims, mine management applied in May 1932 for the existing State Mine Award to be suspended. Justice Beeby rejected this application, suggesting as an alternative that the Victorian Government establish a new, local State Coal Mine Tribunal to deal with wages and conditions. The government readily agreed to do so. A Bill to this effect materialised in mature form at very short notice, raising concerns as to its intent. The Federation protested against the 'many objectionable' provisions of the new Act but accepted its legitimacy. McKenzie supported the new tribunal, accepting the necessity 'to make some reductions in the wages of the men employed in the industry. There is no question, cloak it as we may, that is one of the chief objects of the Bill'.[9]

Encouraged by progress in establishing a separate tribunal for Wonthaggi miners, McLeish moved to reduce the workforce. In four days, between 22 and 26 June, 219 miners lost their jobs. Another 94 joined them on 5 August, and a further 20 on 8 August, with dismissals ceasing only later that month when 430 men had been laid off. To cement a decisive defeat for the union and its members, on 23 August the new tribunal imposed a further general wage reduction of 20 per cent. Although hundreds left the town to face unemployment in Melbourne, where relief seemed easier to obtain, the numbers of unemployed in Wonthaggi still reached more than 500 by October. In dismissing Federation proposals for work-sharing, the Railway Commissioners argued that the inevitable short-time would only lead to 'unrest, discontent and dissatisfaction'. Wonthaggi correctly read this response as self-righteous cant, and the commissioners' tone ensured that, far from defusing 'discontent and dissatisfaction', the sackings served only to further ignite the miners.

While unavoidable in management's eyes, the mass dismissals of June–September 1932 destroyed what remained of Wonthaggi's post-1925 prosperity. Individual miners and their families faced a stark future. One veteran miner, now retrenched, wrote to McLeish:

---

8   Undated memorandum from J McLiesh to the Railway Commissioners, 5 April 1932. State Coal Mine Collection, Public Record Office of Victoria.
9   *VPD*, 1932 Session, 6 July 1932: 545.

I arrived in 1913 and commenced work at once in No 10 on the dog watch as a wheeler, left in 1915 for war where I was twice wounded and returned in 1919... I have worked in all pits and have never been before management for any complaints. Both my parents are buried here, they being amongst the first on the field. I am married with 2 children and have always given my best to my work.

Retrenchment cut a deep swathe through the mining community. John Connelly, son of the VCMA's first president, after being laid off wrote: 'I am a married man and have three children to keep. I have been a householder since 1915 and have never left the mine at any time'. Another miner with intimate links to the State Mine wrote similarly: 'As you are aware of my father being killed in this mine, I am forced to keep my mother, and this dismissal makes it impossible for us to make a livelihood'.[10]

Management was determined to restore economic balance to the State Mine, but miners had absorbed the lessons of hundreds of such individual disasters. When the promised 11-shift fortnight for continuing employees failed to materialise, the dynamics of protest in the town began to change. Although long-time publicists and politicians such as McKenzie and Easton stuck to the script of moral outrage, miners now began to speak a language of industrial and political militancy – a language seldom heard in Wonthaggi for more than a decade. Union committeeman Arthur Asquith reflected this move to the left. Previously a staunch supporter of McVicars and McKenzie, he now spoke of the government's intention to 'wipe Wonthaggi out' and called for 'a great fight to prevent this'.

Wonthaggi's industrial history is best remembered for the successful five-month strike in 1934. But this strike, and its success, can only be understood in the context of the preceding two years, the 21 months between the beginning of the mass sackings in mid 1932 and the commencement of the five-month strike in March 1934. During that period a new strategy calling for militant industrial action, rank-and-file education and political mobilisation would attract increasing support as existing industrial strategies proved ineffective and established union leaders floundered. On 5 September 1932 Wonthaggi miners went on strike, not to resume work for nearly three months, until 23 November that year. But even in the short weeks between the July dismissals and

---

10 C W Baughurst to J McLeish, 28 July 1932; J Connolly to J McLeish, 26 July 1932; T Trewin to J McLeish, 1 August 1932. General Manager's correspondence files, State Coal Mine Collection, Public Record Office of Victoria.

the beginning of strike action in September, the balance of political and industrial life in the town had changed.

In early August, Broome, long in ill-health, died. McLeish now formally assumed the role of general manager, although he had effectively exercised the powers of this position for some time. The miners deplored his appointment, but others offered him enthusiastic support. One un-named councillor was quoted in the local press as arguing that 'Wonthaggi has been in the background and has slipped further back during recent years because it lacked a leader. In John McLeish we have the only man capable of making the mine and Wonthaggi the place it should be'.[11] Wonthaggi would not stay 'in the background' for much longer, but not for the reasons that the *Powlett Express* might have anticipated. McLeish now moved to consolidate his industrial advantage, informing the union of his intention to dismiss inefficient men irrespective of seniority. On this issue the Federation would not be moved. Miners understood seniority as their primary defence against job victimisation: 'members would be retained in order of seniority, the last to come was the first to go'. Rejecting even McLeish's humanitarian gesture of offering to retain some married miners, the union demanded that he instead publicly accept the principle of seniority and abide by it, a demand he formally rejected on 1 September. In the tense discussions that preceded this strike, District Vice-president Idris Williams, another postwar arrival in Wonthaggi, told McLeish, 'we cannot hold the position any longer. I did not think we could hold it as long as we have. We have done our utmost'. With the support of the federal Council, strike action began days later.[12]

There is no doubt that any protagonist in an industrial dispute can obtain an advantage by being a little disingenuous, by reducing analysis or interpretation to over-simplified questions of right or wrong. But in the case of this three-month strike late in 1932, McLeish's repeated hints at darkly inspired external intervention ignored some fundamental truths behind the Federation's position. If he had read, or re-read, the letters received from miners dismissed only months before, he would also have understood these reasons better. It was not simply a strike 'largely influenced by the "Red" element', as he claimed, for while there were 'Reds' in the workforce, they were, at that stage, few in number and hardly influential. What gave

---

11 *Express*, 19 August 1932.
12 See the arguments of both management and union representatives in the unpublished 'Minutes of Conference' dating from August and September 1932. State Coal Mine Collection. Victorian Public Record Office.

the strike, and the union, its force was instead the radicalising effect of depression and mass dismissals on a growing number of mineworkers who had loyally supported an entirely different industrial strategy until only weeks before. Idris Williams is a good example. If McVicars exemplifies Wonthaggi's history of Labor-oriented unionism, Williams stands as the archetypal militant produced by 'Red Wonthaggi' during the Depression. Born in the coal community of Mountain Ash, South Wales, in 1895, he began work at the Powell Duffryn mine at age 13 before serving in the Royal Field Artillery on the Western Front, where he lost a leg in the battle of the Somme in 1916. He emigrated to Australia in 1920 to work at the State Mine. A natural leader, he soon became deeply involved in the social and cultural life of the community as well as the industrial affairs of the Federation. Known as a leading choral singer, Williams also conducted the Wonthaggi Union Band, chaired the Union Theatre Committee and was secretary of the East Wonthaggi Football Club. Increasingly radicalised during the events of 1932 to 1933, he joined the Minority Movement (MM) and ultimately the Communist Party of Australia (CPA). Influential during the 1934 strike as District vice-president, he gained immense local prestige. He served the union as state president between 1934 and 1946, as secretary in 1946–47 and later as national general secretary. He was the first communist to be elected to the Wonthaggi Borough Council, serving a term between 1944 and 1947.[13]

By September 1932 a split had developed within the union, principally between established officials such as Russell, Dowling and McVicars and an increasingly belligerent minority grouped at this early stage around the evergreen militant Jack Goldsmith and Idris Williams (who moved rapidly to the left as confrontation appeared inevitable). Determined on strike action, this group temporarily won majority support, buoyed by the initial reluctance of the union's federal Council to support them. An anonymous report provided to McLeish by a miner present at a union mass meeting early in September provides something of the flavour of these days

> while the National Executive viewed the matter with grave anxiety and concern at the position in Wonthaggi, it regretted that matters would have to stand in abeyance until the meeting of the ACTU… the only suggestion that the Executive could offer was that the Wonthaggi miners should do the best that they can… [In moving a vote of no-

---

13 Phillip Deery. 2002. 'Williams, Idris (1895–1960)', *Australian Dictionary of Biography*, vol. 16. Melbourne: Melbourne University Press: 552–553.

confidence in the Executive] Williams suggested that the letter should be treated with contempt and this suggestion had some support from the floor.[14]

Protection of seniority stood at the core of this strike, but it also proved to be the first occasion at which an alternative log of demands was produced: the strikers called for the introduction of a six-hour working day, a five-day working week, no wage reductions and improved support for the unemployed. The similarity between this set of demands and those advocated by the Communist-led Minority Movement did not pass unnoticed. Shortly after, the National Executive reversed its decision, now endorsing Wonthaggi's call for a wider stoppage. Whatever the reason for such a change, the call failed. Victorian unions pleaded financial distress, while the Sydney Trades and Labour Council interpreted the call as a threat to Jack Lang's attempt to consolidate his hold over left-wing unions. Even other Federation districts rejected the Victorian appeal, to the extent that Northern District members refused to collect levies for Wonthaggi strikers.

In Wonthaggi itself, the dispute reached a deadlock. Union officials continued to plead the cause of seniority to the mine's management but could not budge them. Bereft of external support, the militant wing of the Branch saw the advantage it had held in early September eroding. Mine management waited, confident that neither the miners nor their union could resist McLeish's ultimatum for long. Justice Winneke of the Victorian Coal Tribunal finally broke the impasse in early November when he aired the possibility of a smaller wage reduction in exchange for an immediate end to the strike. In an apparently co-ordinated effort, McLeish simultaneously let it be known that 'management was prepared to deal reasonably with miners if they resumed work'. This apparent compromise proved sufficient. Although the strikers had received widespread support from local businesses (the co-op providentially distributed A£10,000 in dividends during the strike) and strike benefits had been provided to members, the union's financial resources could not sustain indefinite industrial action. By a relatively narrow majority of 293 to 211, members agreed to a resumption, although they still refused to concede the right to maintain employment on the basis of seniority. If not a clear victory, militant miners could claim that management had been fought to a standstill and could also point to minor concessions as a consolation.[15]

---

14   Based on an anonymous undated report of the Federation's 7 September stop-work meeting. State Coal Mine Collection, Victorian Public Record Office.
15   ACSEF Minutes, 18 November 1932.

To many in the community, this November settlement, which returned 1200 miners and 190 ponies to the pits later that month, brought an end to a period of upheaval and instability and a return to industrial peace. But many unionists didn't see it that way and neither did McLeish. In a perceptive note to the Railway Commissioners in the dying days of the strike, he wrote:

> Asquith and Williams are [now] the leaders of a strong minority and they have the support of Goldsmith who can generally command a following. Russell and McVicars are frightened of what might happen and are not strong enough to have the courage of their convictions... [They] were prepared, if possible, to sidestep the seniority question but were not strong enough to impose their views on Asquith and Williams when the matter presented itself for decision in concrete form.[16]

Eighteen months later, with the bitter taste of defeat in his mouth, McLeish returned to this theme, arguing once more that 'much, if not all, the industrial trouble at the mine during the eighteen months prior to the strike was caused by internal dissention in the union, and the disruptive influence of the section comprising the Minority Movement'.[17] Partly true, perhaps, but ironically McLeish's own failure contributed to this process. His attempt to restructure the workforce on his preferred terms had instead directly led to a confirmation of the principle of seniority. As a consequence, the brunt of dismissals fell on those employed after 1928, upon younger mineworkers. Men with sufficient seniority avoided retrenchment; many of whom were the British immigrants of a decade earlier. Seniority had preserved intact the core of British migrants from whom communist and Minority Movement organisation would develop in Wonthaggi.

While McLeish's assessment of McVicars and Russell can be discounted to a degree (they had certainly not delivered the outcome he had expected), the rest of his analysis accurately reflects the consequences of a union branch now deeply divided. He also identified a major change in militant tactics during those crucial 18 months: abandoning mine management as a primary target, militant tactics such as snap strikes and stop-work meetings increasingly sought to put pressure on incumbent officials by forcing them to choose between operational efficiency at the mine and unstinting

---

16 Memorandum from J McLeish to the Railway Commissioners, 21 November 1932. State Coal Mine Collection, Victorian Public Record Office.

17 J McLeish, 'The State Coal Mine's Industrial Situation'. Report to the Railway Commissioners, n.d. (July–August 1934). State Coal Mine Collection. Victorian Public Record Office.

support for every grievance the mineworkers had. But McLeish's view, however perceptive, is deficient. He underestimated the catalytic effects of the Depression, not only within Wonthaggi but on a far wider social canvas. In terms of the State Mine workforce, McLeish also exaggerated the organisational maturity and influence of the Communist-led Minority Movement during 1932 and 1933. Both need to be considered in setting the scene for the 1934 strike.

As a community dependent on a single staple industry, Wonthaggi relied absolutely on the health and future of the State Mine. During these years, however, standard measures of security disappeared under the impact of mass retrenchments and wage reductions. Coal production collapsed as unemployment dramatically increased. This threatened the State Mine's operations and the town's future. Under such circumstances rumours thrived, such as the story that ran like wildfire through the town in late February of a further 300 dismissals. McLeish's understandable reluctance to confirm or deny every whispered conspiracy only served to lend them temporary credence. In an attempt to dispel such fears and restore a measure of community confidence, a public conference considering the future of the State Mine convened in February 1933, attended by McKenzie and by Victoria's ambitious Attorney-General Robert Menzies. The two provide a study in contrast. Menzies did little to reassure miners, emphasising instead the precarious condition of the mine (a theme he had been developing for more than a year), while McKenzie continued to doggedly advocate lower prices for State Mine coal and a resumption of public coal sales as a panacea. In his eyes, only an ambitious, expansionary program could resolve local economic, and hence social, problems. Although willing to acknowledge the mine's indirect advantages to the state's economy, Menzies proved reluctant to concede McKenzie's arguments, particularly any suggestion of an expansionary investment at the mine. Menzies offered instead an austere solution that, under the circumstances, hardly sounded reassuring: he 'sought to secure for a reasonably constant body of men reasonably constant employment at a reasonable rate'. At this conference, Menzies performed as an uncompromising politician, delivering a blunt message to the miners that conditions would necessarily have to be further reduced to protect existing jobs and, moreover, that resistance by mineworkers could only endanger the State Mine's future. [18]

---

18 The best report of this meeting is probably that published in the *Sentinel*, 24 February 1933.

But the time for conferences had passed, or perhaps had yet to come. The trial of strength within the union, and between miners and management, proved too urgent. In response, the union campaigned to lay aside the arrangements whereby they could be required to work an additional shift each fortnight for a nominal sum (at times only 6 pence) in favour of a standard 10-shift fortnight attracting a uniform basic wage. McLeish retaliated by seeking an increase in hours of work. Though ultimately not pursued before the tribunal, his application served to kill the miners' 10-shift campaign almost before it had begun. Such cat-and-mouse games continued sporadically throughout 1933. In the 10 months to December 1933 a further 27 days were lost to stop-work meetings as militant members of the union stepped up pressure within the union and against management. They knew that concern over industrial issues could win an audience, and issues as diverse as cavil rules, unfair dismissals, the use of ponies, the 10-shift fortnight, differential wage rates, the behaviour of the tribunal, and wheelers' working hours were considered at these meetings. But behind these niggling actions an essential truth remained: the union's work-sharing schemes had been rejected and unemployment soared.

As organised resistance grew within the Federation, so too did the influence of retrenched miners within the Wonthaggi Unemployed Workers Union (WUWU). In contrast to many of the town's earlier unemployed, whose unionism had been marginal to their employment, these were schooled unionists, seeking to retain links with the Federation even while unemployed. In this they received every encouragement from the Federation itself, for their continuing acceptance of union leadership not only served to defend the principle of seniority as the basis for employment, but also ensured that there could be little threat of locally recruited 'free labour' being used against those still employed at the State Mine. The Federation assisted in other ways – assistance with registering for state relief, donations to the WUWU, access to union mass meetings, the right of unemployed miners to vote in union elections (although not to stand for office), and the distribution of free copies of the union's paper *Common Cause*. Unemployed miners remained well within the orbit of the Federation.

General attempts to relieve social distress also intensified. Children of unemployed families were now fed daily at school by the town's Public Assistance Committee, which also organised the distribution of shoes, boots, clothing and other necessities. Even so, charity had its limits. A call by the WUWU to allow the unemployed to 'work off' their rates through community work failed, for 'it was not a business proposition'. Yet calls such as this one did have some effect, as the payment of rates in Wonthaggi could, and did, remain outstanding for years. Families renting accommodation proved far more vulnerable. Although unemployed families in Melbourne facing eviction could at least apply to be granted cheap accommodation, every effort to extend this principle to regional Victoria had been rejected. While the inadequacy of the 'susso', the state-funded sustenance relief, attracted much condemnation, the lack of accommodation proved the real litmus test. Everyone recognised that a family evicted in Wonthaggi must necessarily leave the town, as no other accommodation would be offered. Only one alternative existed. Generally families were loath to undertake this extreme step, but an unknown number did so in order to remain close to the town and possible future employment at the State Mine. Families divided and broke up. Children and parents lived separately with relatives and friends, in fact with anyone with a home and a job. In a community where unemployment stood at more than 20 per cent, even following mass departures, mineworkers and the unemployed shared a desperate search for security and survival.

And what of McLeish's belief that after September 1932 a communist minority effectively dictated terms in the local Federation Branch, making a major strike inevitable? McLeish argued that increased disruption in the 18 months prior to the 1934 strike reflected an internal union struggle between a moderate leadership and a disruptive minority acting in the interests of the communist-led MM.

The reality was not so simple. Some issues are clear: first, that militant unionists now sought to defeat established officials; and second, that by late 1932 a communist-led faction opposed to McVicars and Russell existed within the Branch. By 1934 both the MM and the Communist Party were well established among the State Mine's workforce. But despite a number of suggestions, no permanent MM group operated at the State Mine before mid 1933 at the earliest. The year 1933 proved to be pivotal: moderates lost their authority and a mood for strike action gained ground within the Branch. The MM had devoted considerable efforts to expanding its influence in the mining industry but found its first successes on NSW mining fields, such as when Bill Orr, a communist and MM leader from the NSW Western District, was elected general secretary of the Federation in January 1934. In Wonthaggi, union politics now proved equally volatile, marked by shifting allegiances and changing political fortunes. 'Bob' Russell, seen as a moderate District president had, in fact, resigned from the ALP in June 1933, while one of the principal leaders of the militant tendency, Idris

Williams, remained within the ALP, or apparently so. Biographical details published by the Federation in later years suggest that he may have secretly joined the Communist Party in late 1932 or early 1933. If true, such a decision only serves to highlight the uncertainties and manoeuvrings that marked Federation politics during these depression years. There had been individual members of the Communist Party among the mine's workforce since at least 1930, but no organised party branch until 1933. Irrespective of the allegiance, public or private, of union members such as Williams, before late 1933 the Communist Party and the MM lacked industrial authority in Wonthaggi. Similarly, the industrial claims of the MM were undoubtedly familiar to many in Wonthaggi, but this alone could not translate into either an effective MM unit before 1933 or control of the Victorian District until the following year. Only in January 1934, when confrontation appeared inevitable, did the MM begin to publish a local pit paper, *The Sprag*. It ran for nine issues, until superseded during the 1934 strike by a more ambitious paper, *The Miners' Voice*. *The Sprag* took its title from mining terminology, referring to a solid lump of wood or iron placed between the spokes of skip wheels, which jammed against the bottom of the skip to prevent it from travelling, an appropriate metaphor for militant miners. From the ranks of the MM were drawn the men who would lead the district and occupy federal office during the next 20 years. The ideological tone and general strategy of the MM paralleled that of its British counterpart, and leading activists were overwhelmingly British, men for whom the experiences of war as well as the demoralising defeats of the Miners Federation of Great Britain in 1921 and again in 1926 exercised a profound influence on their political and industrial attitudes.

What changed? Firstly, the Labor Party lost much of its political authority and with it the allegiance of many active unionists. Electoral disasters in both federal and state politics compounded the problems of Labor-affiliated union officials now struggling to effectively defend members' interests. Similarly, arbitration had lost its potency, and was now seen by many as an instrument to depress working conditions, not to protect them.

During late 1933 two events served to bring industrial matters to a head. In October, former prime minister and arch union-hater Stanley Melbourne Bruce resigned from the federal seat of Flinders. The Communist Party resolved to contest the by-election, and campaigning concentrated on Wonthaggi. A Party branch had been established some months earlier, developing out of a socialist debating group established in 1930 to discuss current political and social questions. It would be another six months before the Communist Party would finally abandon its position of outright hostility toward the ALP and the Party press still spoke heroically of its reception in Wonthaggi: 'strong handgrips and brave greetings followed by curses against the system… fierce denunciations of the traitorous politicians and trade union officials who had betrayed them in their past struggles'.[19] The Communist Party put forward the mildly spoken university graduate Ralph Gibson who polled respectably, winning 3100 primary votes (or 5 per cent of the poll), more than sufficient to encourage further campaigning.

At the same time, McLeish decided to reassert his authority. Holding firm views on the reasons behind the constant interruptions to State Mine production, the manoeuvrings within the union must have driven him to distraction. Each lost day cost the mine invaluable revenue and, in his eyes, represented a challenge to his authority. He responded a fortnight before Christmas with an ultimatum: miners 'had to decide whether [they] were going to obey the instructions of the officers of the mine and carry out orders given' or face dismissal. For good measure, he once again threatened miners failing to 'make the minimum' (in other words, to meet required production targets) with suspension. In the town's overheated industrial environment, many interpreted his ultimatum as a further threat to wages and conditions.[20] These issues festered over the New Year shutdown, and by late January it had become apparent that the union's militant wing saw in McLeish's demands an opportunity to advance its agenda. Reports from Wonthaggi had by now become a regular feature in the MM's paper, *The Red Leader*, which in late January 1934 carried a particularly explicit appeal to arms

> [an] earnest appeal to our fellow-workers in the mining industry to back us up in the coming fight against further reductions in wages and worsening of conditions… The miners of Wonthaggi will always put up a fight if given the right leadership. We think that the next twelve months are going to be a time of great struggle… [21]

A 'great struggle' finally seemed inevitable. By February 1934, after nearly two years of manoeuvring, it appeared that all parties now welcomed a confrontation. McLeish sought an end to uncertainty and a clear assertion

---

19 Quoted in *The Workers' Voice*, 1 December 1933.
20 His demands, and the union's initial response, are summarised in the ACSEF Minutes, 14 December 1933.
21 *Red Leader*, 24 January 1934.

of managerial prerogative. McVicars and his allies sought to escape the impossible position of acting as increasingly unpopular negotiators between militants and managers; while Williams, Asquith and their MM supporters believed that a successful strike would vindicate their tactics to win power and control of both the Victorian District and Powlett River Branch of the Federation. A wild card was Jimmy Jones, an MM organiser who had arrived in Wonthaggi in October the previous year. A solitary, nomadic rebel from the Northern coalfields of NSW, Jones claimed either Welsh or English nationality as the occasion demanded. His organising skills proved instrumental in developing a core MM membership of 30 to 40 miners by early 1934. In the rough-and-tumble of public debate, Jones could be pugnacious and, if necessary, insultingly rude. On the very eve of the strike he managed to infiltrate a union mass meeting, one at which McVicars sought to impose his authority. Spotting Jones in the crowd, McVicars called on him to defend the MM and its strategy. Jones refused point-blank, telling McVicars instead that he 'was not the champion of the [MM]. Mr William Orr was the champion, and [Orr] would tear McVicars to pieces'.[22] By this time no-one could precisely calculate the relative popularity, or acceptance, of alternate industrial strategies. Support could change from meeting to meeting. Mass meetings in early March supported the decision to oppose what were seen as management attempts to once again undermine wages and conditions, but an attack by Williams and his allies on the legitimacy of the Coal Tribunal met with defeat. Summoning all his resources, at one of these meetings McVicars successfully moved a motion of confidence in the Executive's decision to negotiate with the tribunal if necessary, while simultaneously condemning reports of Wonthaggi negotiations carried by the MM's paper, *The Red Leader*. Clearly, neither faction had a decisive advantage in terms of numbers or arguments.

Perhaps McVicars's limited success encouraged McLeish, or more likely McLeish was following his own timetable. Whatever the case, his summary dismissal on 5 March of seven men for inefficiency and a further two for insubordination triggered general strike action. In the face of such a direct threat to its own authority, the Federation closed ranks. McLeish added insult to injury by dismissing more miners working in the Dudley Area pits on the following day. A total stoppage immediately followed. No-one, least of all McVicars, now sought compromise. A stop-work meeting

---

22 For this entertaining exchange, see ACSEF Minutes, 5 February 1934.

unanimously agreed to stay out until 'an accumulation of attacks, pin-pricking, intimidation and the failure of the management to co-operate in easing the situation, or helping to overcome misunderstandings' had been resolved.[23] With all sections of the union agreed on strike action, the conciliatory tone of the resolution reflects the union's decision to occupy the high moral ground from the outset. But if an apparent willingness to negotiate was deemed necessary to avoid offending public opinion, then the industrial claims made on management shortly afterward displayed a more uncompromising mood. As McLeish moved the mine to a care and maintenance regime, with pit ponies withdrawn and rolling stock redeployed, he received the Federation's own demands. Not only should all dismissals be reversed, but the union now required management's agreement to a number of other reforms, particularly the recognition of union-elected pit committees. More than any other demand, the call for recognition of pit committees reflected the increasing influence of communists and their allies within the Branch; and the centrality of the call for recognition demonstrated the extent to which momentum was swinging its way. Pit committees were designed to extend the reach of the union's authority right to the coalface, enabling elected rank-and-file members to police all union decisions and deal with the myriad job grievances that multiplied underground. The union argued that such committees would reduce disputes. Not surprisingly, McLeish vigorously disagreed, denouncing the proposal as 'a mischievous interference with the functions of management [which] would aggravate petty disputes'.[24]

A war of words erupted as both sides put their case to the public, with the *Case for the Union* rapidly answered by the commissioners' *Wonthaggi Coal Mine Dispute*. The *Case for the Union* pulled no punches:

> The present dispute at the Wonthaggi State Mine is the culmination of a whole series of pin pricking actions and attacks upon the miners conditions by the management over a period of many months. In the mad drive for what the management graciously calls 'efficiency' and 'continuity' of work, they have introduced altered conditions and intimidation to the point where the position of the employees has become untenable.

---

23   Victorian District ACSEF. 1934. *The Case for the Union*. Wonthaggi: Victorian District ACSEF.
24   Memorandum from McLeish to the Railway Commissioners, 2 July 1934. State Coal Mine Collection. Victorian Public Record Office.

Management, the union argued, pursued 'a policy calculated to compel capitulation and unconditional surrender of the union to the management'.[25] But words were one thing, public support another. Both sides knew that community support would prove essential to their success. On 12 March, a public meeting of more than 3000 listened as loudspeakers relayed speeches to both the Union and Plaza theatres and overwhelmingly voted to support the striking miners. Significantly, John Short, manager of the co-op, moved the necessary motion of support, seconded by the manager of the Malvern Star Garage, Ed Vistarini. On an unprecedented scale, Wonthaggi embraced a miners' strike as its own.

Welcome though it may have been, the Federation faced a major task in translating such support into effective resistance. Moreover, simmering disputes between moderates and militants had yet to be totally resolved. During the strike's opening weeks, McVicars and Russell assumed control of the strike, but once again came under fire for their 'wait-and-see' tactics. As the MM pit paper *The Sprag* put it, 'the way to defeat McLeish and Co is not by compiling fine speeches and clever arguments, but plain organisation of food [and] organisation of relief and support'.[26] Militants argued for a new way of running the strike, proposing a separation of authority, with control of strike strategy, relief and propaganda vested in a Broad Committee, a body solely concerned with day-to-day aspects of strike organisation, distinct from the existing union Executive and committee. With McLeish refusing to accept the introduction of pit committees, this argument succeeded with remarkably little dissent. Elections to the Broad Committee provided a foretaste of the District elections held later in June. Together with their sympathisers, militants won the committee's crucial positions. Asquith, now an enthusiastic proponent of the MM position, became president, supported by Tom Currie. A popular and capable miner who had helped form the local Communist Party branch the previous year, Currie belonged to Wonthaggi's generation of postwar British migrants, as did at least three others elected to the Broad Committee: Idris Williams, Bill Stirton and Joe Foster, also editor of *The Sprag*.

---

25 The union's position is fully summarised in Victorian District ACSEF. 1934. *The Case for the Union*. Wonthaggi: Victorian District ACSEF. The management's case appeared in Victorian Railway Commissioners. 1934. *Wonthaggi Coal Mine Dispute*. Melbourne: Victorian Railway Commissioners.
26 The following account of the 1934 strike draws on local newspaper sources, ACSEF Minutes and on Peter Cochrane. 1973. 'The Wonthaggi coal strike, 1934'. B.A. Honours thesis. Melbourne: La Trobe University. Also, Andrew Reeves. 1977. 'Industrial men: Miners and politics in Wonthaggi, 1909–68'. M.A. thesis. Melbourne: La Trobe University.

With Broad Committee activities organised into three subcommittees – Propaganda, Relief and Distribution, and Entertainment – each supported by a team of volunteers drawn from the union's ranks and the wider community, this dispute became known as the 'best organised strike in history'. Given the influence of the British miners, it is not surprising that the Broad Committee resembled, and in all probability drew upon, the local strike committees formed by British miners during the coalminers' long but unsuccessful strike of 1926. The novel combination of co-operative community organisation, mobilisation and coordination of relief efforts – supported by clever and imaginative publicity efforts – in Wonthaggi and also further afield, provided a blueprint seized on by other unions and communities in coming years, and in retrospect can be seen as one of the critical moments in the recovery of Australian unionism from the impact of depression.

The Federation organised relief in Wonthaggi on a communal basis, centred on the Union Theatre. Strike pay remained low, so strikers relied heavily on donated food and clothing. Donations of food were sought from nearby farms to supplement support from the town. Both the Caledonian and Powlett hotels donated loaves of bread, while other local businesses provided fruit and vegetables. Local farmers and businessmen contributed generously, while the Co-operative Society extended thousands of pounds in credit as well as financial donations of its own. Distributions occurred twice-weekly at the Union Theatre, with the queue snaking as much as 100 yards through the theatre and down Graham Street. After initial teething problems, by April and May Relief Committee distributions supported up to 700 miners' families, and while the quantities remained rationed, the variety of food continue to improve. For example, on 10 May the Relief subcommittee distributed 5000 pounds of meat, 1100 loaves of bread, 3 tons of potatoes, half a ton of onions, 750 pounds of dripping, 840 tins of jam and 30 bags of apples. Such relief efforts sustained the strike during its initial weeks. Not until 13 April would the first distribution of strike pay take place, and that depended heavily on local donations supplemented by an initial grant of A£1000 from the Federation and a loan of A£1500 from the co-op. Federation contributions and pledges from other unions meant that from April onward regular strike pay supplemented relief efforts.

Community-based relief expanded during the course of the strike. The Union Theatre stood as an oasis of relief, respite and enjoyment during months of unrelieved tension. Theatricals, movies, talent quests and dancing competitions restored morale and provided a few laughs for miners and their families, now reduced to a minimal standard of living. Picture shows screened nightly at the Union Theatre, and even with admission prices heavily reduced, these shows proved an important source of revenue for the Broad Committee's relief efforts.

It was clear such a level of organisation would be necessary if miners were to avoid outright defeat. With adequate coal reserves and uninterrupted supplies from NSW, the Railway Commissioners had every expectation that the strike could be confined to Wonthaggi. For the union, marginalisation had to be avoided at all costs. The Propaganda Committee organised a national speaking tour: a team of 12 miners and their wives descended on Melbourne; another five concentrated on Victoria's Murray Valley; while others ranged as far afield as Tasmania, South Australia, Western Australia and Broken Hill. In the first days of July, at the Northcote Town Hall in inner Melbourne, Agnes Chambers spoke to a packed meeting with her powerful message directly addressed to other working women:

> Do the women of Australia understand what the life of a coal-miner is, how dangerous is the occupation? We women, living at home, while our husbands are toiling in the bowels of the earth each year, see the accident lists growing greater and greater each day. When they leave for work, we are not sure whether they will return uninjured. These accidents, our husbands tell us, are largely due to the system of 'speeding up'. Because our men are determined to call a halt to this sacrifice of life and limb, the Government threatens to take from us our homes.[27]

The strikers also enjoyed the active support of the Federation's recently elected general secretary, Bill Orr. Like so many other prominent Australian union officials of these inter-war years, Orr had migrated to Australia from Scotland following war service and disillusionment with his prospects in Britain's postwar mining industry. Orr rose to prominence at Lithgow in NSW as an activist in the MM and the Unemployed Workers Movement, and the Wonthaggi strike offered him an early opportunity to put his own strategies for the Federation into practice following his election. Whereas other Federation districts had shunned Wonthaggi in 1932, the intervening years had changed the alignment and the outlook of the union nationally. Orr acknowledged the specific, localised nature of the strike but shared with the strikers themselves a determination to extend its impact. After

---

27   Quoted in Peter Cochrane. 1973. 'The Wonthaggi coal strike, 1934'. B.A. Honours thesis. Melbourne: La Trobe University.

initial concerns at the capacity of the Federation to wholeheartedly support Wonthaggi's strikers, Orr soon recognised the potential of a successful strike to revitalise the entire union. Perhaps the flying visit of Williams to Sydney in March to consult with Orr had changed his mind. In any case, shortly after that visit the Federation's National Council struck a 2.5 per cent levy in support of Wonthaggi strikers.

Between May and July the strike passed through two expansionary stages. Shortly after successful May Day demonstrations in Melbourne, Orr sought to extend the strike in coordination with the Melbourne Trades Hall Council and the state ALP, targeting 'attacks on the trade union rights of Australian workers'. By seeking to embarrass the Victorian Argyle Government and the Railway Commissioners, Orr hoped to develop sufficient momentum for a campaign to reverse the earlier 10 per cent wage cut suffered by Federation members and other unionists. This strategy led to a crisis in Wonthaggi, where unity had been firmly based on the defence of threatened local conditions. Concerns were now publicly expressed by some borough councillors and local businessmen, as both groups increasingly sought to bring the strike to some sort of satisfactory settlement. Concern, too, was expressed at the growth of the MM throughout the strike. By May, with more than 200 miners active in strike organisation, the MM had established itself as the leading force within the local Federation Branch. By the end of the strike it could claim 300 members, a claim confirmed by the results of the bitterly contested June union elections. Among the moderates, or 'constitutionalists', only McVicars successfully defended his executive position. Williams won the presidency and MM activist and Broad Committee member Bill Stirton became vice-president, with Asquith as Federation Delegate. More than half of the 12 committee positions also fell to militants.

This election campaign was enlivened by the circulation of a fake edition of *The Sprag*. This fake edition (of a pit paper that had ceased publication two months previously) bitterly attacked Williams, variously referring to him as a 'rat' and as 'vermin'. Miners identified it as having been printed on a roneo machine in the State Mine's office, an identification made possible by the pin hole that this particular machine left in the top right-hand corner of each sheet that it printed. *The Miners' Voice* called for calm: 'this fraudulent document is intended to provoke disturbances. Should this occur, the town will be flooded with police and the way cleared for scabs'. Union esteem for the mine's management could hardly fall any lower. In any event, the fake *Sprag* seemingly influenced no-one; miners held their peace, no extra police arrived and militants took control of the Victorian District.

The approach to Melbourne unions and the Trades Hall Council coincided with a major expansion of the local relief effort. In early April the Relief Committee established its own slaughter yard to process cattle purchased directly from farmers, and arranged bulk purchases of potatoes, onions, vegetables, jam, bread and tea from the co-op. Teams also travelled the 40 miles to Pakenham where fruit from the local orchards could be obtained at a nominal cost. A free barber's shop now operated in the union's rooms and a team of seven union members established a boot repair operation, collecting and repairing more than 1500 pairs of boots during the course of the strike. By May, with such arrangements in place, relief efforts had been extended to other unionists in the town, particularly the coal carters who had been hit severely by the strike.

Notwithstanding the increasing concerns of local businessmen at the long-term consequences of the strike, support for an extension to strike action grew. In early June the *Powlett Express* confidently reported: 'it is safe to say that 90 out of every 100 in Wonthaggi are of the opinion that an all-out policy is the only way of bringing this deadlock to a successful conclusion'.[28] Orr sought to build on this success nationally, and on 18 June the union's federal Executive endorsed the slogan 'Prepare for a General Strike', declaring that 'victory for Wonthaggi and for the rest of the trade union movement demands that the present campaign shall be extended as widely as possible'. Was this a bluff? Perhaps so, but it worked. The threat of such a strike, real or imagined, would never be tested, as the Victorian Minister for Railways and government spokesman Robert Menzies executed a sudden policy reversal. Intransigent since March and loath to believe the strike to be more than an irresponsible disruption, he had by early July offered not only to 'favourably' consider reinstatement of dismissed miners but also to recognise pit committees if the Federation immediately returned to work. Although pit committees had been a bone of contention during the preliminaries to the strike, miner Bill Stirton recalled how relatively simple the solution had become by July:

> In that period there were numerous instances of men going to work and some form of dispute… would arise and as there was no machinery whereby a discussion re the dispute [could occur] all the men would come home without going to work. This sort of organisation was not good for the union and caused a fair amount of disunity. It was decided that if we [the union] could have a pit committee consisting of two

---

28 *Powlett Express*, 4 June 1934.

Figure 22. Food preparation and distribution during the 1934 strike: the butcher's team preparing meat for distribution
Wonthaggi Historical Society Collection

Figure 23. Mineworkers receiving pay during the five-month Wonthaggi strike
Secretary Jack McVicars on the right doles out relief funds while leading militant Bob 'Hammie' Hamilton (third from right) keeps a keen eye on proceedings.
CFMEU Mining and Energy Division National Office

members at each pit on each shift to consult with the undermanager and another official most of these unnecessary stoppages could be resolved. The management absolutely refused to have anything to do with pit committees. This was one of the points raised [during July] at the conferences between the management and the union at conferences called by the Victorian Minister for Railways, Mr R G Menzies. Mr Orr stressed this point as did other union representatives and Mr Menzies said that he could not see anything wrong with this proposal. The formation of pit committees was one of the terms of settlement.[29]

Faced with potential industrial turmoil on a scale sufficient to reduce the Wonthaggi strike to marginal interest, acceptance of the strikers' demands must have seemed a preferable, if for some a humiliating, alternative. For the Federation, and its Wonthaggi members in particular, the election of pit committees on 17 July and the subsequent return to work on 20 July represented success on a scale far larger than simply the commissioners' abandonment of threatened wage cuts and further dismissals. 'Defend the Wonthaggi miners' had proved to be a potent slogan around which the Federation's districts had regrouped. Under militant leadership, a now-united Miners Federation moved to recover lost wages and working conditions. Out of Wonthaggi's victory developed Federation-wide success later that decade.

For Wonthaggi and its mining workforce, union success in 1934 reshaped their future. When in February of that year McLeish had intimated private support for a 20 per cent wage reduction, he had been motivated by neither antagonism nor irresponsible provocation. He aimed to make the State Mine a profitable state enterprise once more, capable of re-investing in its own capital needs and returning a dividend to the state. Miners, he believed, must submit like others to the consequences of the Depression. His reasons for concern are readily evident. The Depression had cut a swathe through the smaller collieries and mines in the Gippsland region. Locally, the mines at Kilcunda, along the Bass coast from Wonthaggi, and at Jumbunna, to the north, had closed. McLeish would also have watched the State Mine itself decline in importance. In 1929–30 its output represented more than 6 per cent of Australia's total production, a percentage that by 1935 had dropped to barely 3.5 per cent. McLeish feared for its survival, believing that a minor colliery on a peripheral coalfield could not long survive. His efforts had failed to restore lost profitability. In the words of Robert Lee, who conducted an inquiry into the operations of the mine following the 1934 strike: 'it must

---

29   Bill Stirton. 'Depression in Victoria', mss notes provided to the author, 1974.

now be decided whether [the mine] can bring to the State such benefits as will fully compensate for the losses which must be faced'.[30]

Paradoxically, Wonthaggi's political importance increased as its economic significance continued to decline; 1934 had proved to be a watershed. The resistance of State Mine employees provided a model for the Federation's post-depression recovery. Equally importantly, methods of industrial and community mobilisation refined in Wonthaggi were adopted by the Federation and other coal communities, forming the basis for the successful national industrial campaigns of 1937 to 1940. Wonthaggi's strike leaders, especially Idris Williams, emerged as leading members of the union at the national level. Williams would eventually move beyond Wonthaggi, called to Sydney to serve as the Federation's general secretary in 1947. Ironically, it would be Williams, one of the victors of 1934, who would lead the Federation into its climactic postwar confrontation, the disastrous 1949 Miners' Strike.

---

30   Robert Lee. 1934. *Report on the Victorian State Coal Mine, Wonthaggi*. Melbourne: Railway Commissioners: 14.

Chapter 4

Political recovery and economic decline, 1934–1939

Residents returning to Wonthaggi in 1935 after a decade's absence would have had no difficulty in quickly getting their bearings. The mine's central operations still dominated Wonthaggi's western approaches and its working pits circled the town. From the Union Theatre and Workingmen's Club in the east to the Union Dispensary and the Co-operative Store at the far end, Graham Street had retained its familiar, commercial air. But behind the commercial facades of Graham and McBride streets, and in the surrounding miners' cottages, the community had changed. Commerce recovered more slowly from the effects of the Depression than the miners' union, as the 1934 strike emphatically demonstrated. If pre-Depression public campaigns in Wonthaggi had concentrated on possibilities of mine expansion, then post-1934 campaigning centred on the preservation of jobs and the survival of the mine. In such campaigning the Federation now naturally assumed the lead.

This was not easy, for throughout the 1930s the Australian coal industry stood at odds with the general drift of the economy. Even as late as 1937, aggregate production fell nearly two million tons short of equivalent figures from 1924. Although production did pick up after the absolute depths of 1931, the cautious rate of recovery guaranteed neither jobs nor security. In the opinion of the Federation's new, left-wing leadership, tentative recovery meant the threat of increased mechanisation in the mines and structural unemployment for mineworkers. Others shared this view, even if conclusions differed. As early as 1929, the influential economist F R E Mauldon had warned of the dangers of ignoring the need to rationalise the costs of production and marketing, expenses that he suggested drove up costs for the entire industry. He argued for the closure of many smaller mines and the loss of up to 5000 jobs in the industry.[1]

---

1 F R E Mauldon. 1929. *The Economics of Australian Coal*. Melbourne: Melbourne University Press: 85–86.

Figure 24. Members of the union pose underground during the early 1930s to demonstrate the inadequacy of working conditions at the State Mine
McVicars, District Secretary (wearing a tie), sits on the left among his members.
Wonthaggi Historical Society Collection

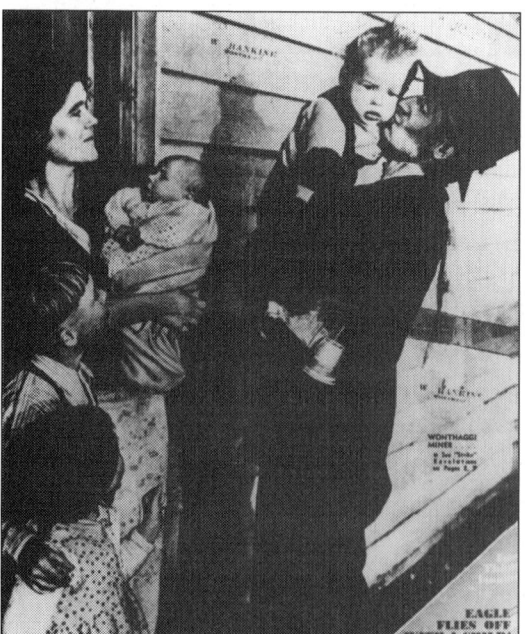

Figure 25. An image of Depression-time mining communities that could be replicated across the country
Gaunt faces, raw weatherboard houses and patched clothing. Full employment only slowly returned to the industry.
Wonthaggi Historical Society Collection

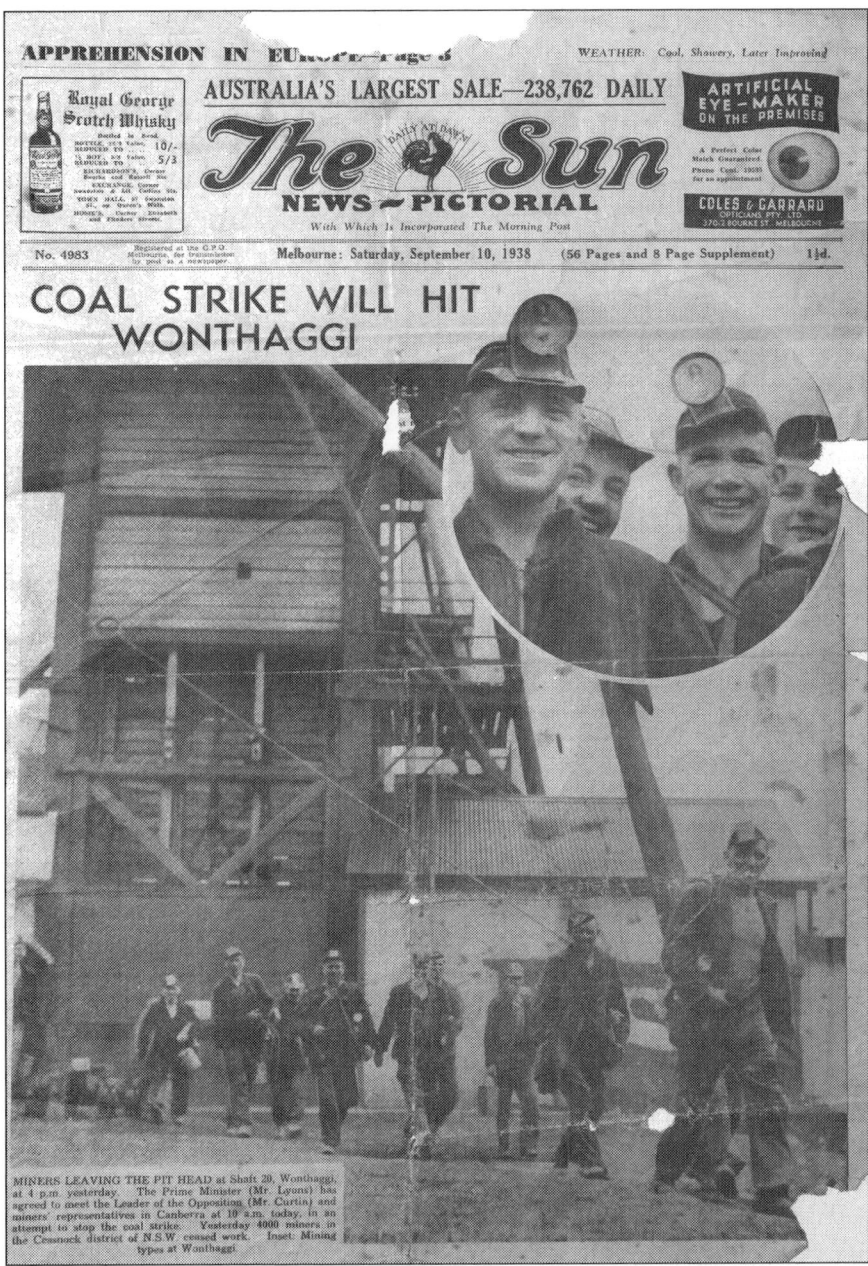

Figure 26. Front page of the Melbourne *Sun News-Pictorial*, 10 September 1938, showing Wonthaggi miners striking in support of the Federation's national industrial claims
Wonthaggi Historical Society Collection

The Depression vindicated his argument. But the Federation offered different answers to the problems created by 'lazy coal producers' (Mauldon's own description) and, after 1934, remained eager to not only protect mining communities but to rebuild employment in the industry as well. In seeking to build upon its victory at Wonthaggi, the Federation now produced its own analysis of the state of the industry and a blueprint for economic recovery. Between 1935 and 1937 Bill Orr, with assistance from the national president Charlie Nelson, outlined the union's case in a famous trilogy of pamphlets: *Coal: The Struggle of the Mineworker*; *Mechanisation, Threatened Catastrophe for the Coalfields*; and *Coal Facts*. The scale of the union's concern and something of its proposed response can be seen in the opening lines of *Mechanisation*:

> more than one hundred thousand industrial workers, small farmers, business people and professional workers, who are directly or indirectly dependent on the coal industry face ruination and unemployment if the coal owners' plans for further rationalisation are allowed to mature.[2]

From a union perspective, Orr and Nelson painted a grim picture. Rationalisation and accompanying job losses stood at the heart of their concern (and their response) but, as they wrote in *Coal*, 'the methods used by the coal bosses to achieve their ends are many and varied'. They produced an impressive list of immediate threats to mineworkers' livelihoods: a general 'speed-up' of work across the industry, the threat to close individual mines unless production targets were met, the destruction of miners' self-imposed limits on production to protect miners' jobs and health, and attempts to eliminate concessional payments for wet or otherwise deficient working places. The Federation deliberately chose to fight on two fronts, arguing for an economic alternative to mine rationalisation while also advancing a case for social preservation: to protect individual jobs, smaller mines and above all the dozens of communities that depended on coal for their livelihood. Inevitably, these two strategies entwined: 'mechanisation can and must be resisted when it threatens the very existence of whole communities who are denied any alternative means of earning their living'.[3]

In an established colliery, three major variables determine production: sufficient exploratory and development work to open new seams, the presence of sufficient miners to work them, and a market sufficient to

---

2 W Orr. 1935. *Mechanisation, Threatened Catastrophe for Coalfields*. Sydney: Miners Federation: 3.
3 W Orr and C Nelson. 1935. *Coal: The Struggle of the Mineworkers*. Sydney: Miners Federation: 9.

consume the coal produced. Many believed that the miners' willingness – or unwillingness – to work was the primary cause of the State Mine's troubles, and it is true that strike action, stop-work meetings and other union action did interrupt production. But by 1935 such an argument could no longer stand unchallenged. By that time national coal production had started to increase, albeit slowly, yet production at the State Mine continued to decline. Even if strike action did continue to punctuate the operations of the mine, it alone could no longer be blamed for falling production, stagnant employment and lower productivity. Even worse, the growth in national production reflected an increase not of coal exports but of domestic consumption, the very market the State Coal Mine had been established to supply.

In framing retrenchments in July 1932, McLeish had produced a nine-year development program for the mine in order to maintain annual production of 500,000 tons. McLeish anticipated the exhaustion of established operations in the McBride Tunnel as well as in the Dudley, Northern and Eastern areas by 1937, offset by the progressive development of the Western Area after 1933 and the Kirrak Basin to the east after 1935. These two areas were scheduled to both reach optimum production of 900 tons daily by 1938. Depression, lack of government finance and unfavourable geology first delayed and then broke this finely tuned plan. Only in the aftermath of the 1934 strike did co-ordinated developmental work on the Western Area commence, 'too late', McLeish commented, 'to arrest the decline in output'. In successive Annual Reports throughout the remainder of the decade, he continued to argue for the resources and approval to push ahead with work in the Kirrak Basin, but with limited success. Although scheduled to produce 900 tons daily by 1938, Kirrak was, as late as 1941, struggling to produce 100 tons. By late 1936 McKenzie claimed developmental work to be 'two to three years' behind where it should be; by 1940 that delay had increased to at least four years. In 1936, McLeish had to rework his developmental program, now anticipating that production would peak at 502,000 tons in 1943. Actual figures demonstrate the scale of the challenge he faced: 282,000 tons in 1941–42; 290,000 tons in 1942–43; and 234,000 tons in 1943–44. By then, underdevelopment and falling production had become institutionalised at the State Mine.[4]

Mineworkers, management, local businessmen and politicians all recognised the symptoms of this creeping paralysis. Once more the issue

---

4   For accounts of McLeish's developmental efforts, see *Annual Reports of the General Manager of the State Coal Mine*, 1934–35 to 1941–42. Melbourne: Government Printer.

of mine development, or the lack of it, assumed a political edge, spurred on by the sense of instability and impermanence that the Depression had caused. Although accepting McKenzie's argument that 'rehabilitation of the industry and unemployment are the two main issues confronting us at the moment', Federation members remained unwilling to trade away industrial conditions and renewed wage demands in exchange for investment in mine redevelopment. During 1935 the Federation actually stopped shaft-sinking and other developmental work in protest against what it saw as attacks on wider working conditions. In the Federation's eyes, mine development and improvements to wages represented two sides of the same coin. Neither would, or could, prevail at the expense of the other. The Federation's strongly held position that developmental work, however pressing, could not be considered independently of day-to-day industrial relations, heightened political tensions in the town. McKenzie rejected both the Federation's analysis and its actions. In his estimation, mine development must proceed at all costs. Having succeeded in attracting about A£180,000 in developmental capital from the Argyle state government in the wake of the 1934 strike, he sought absolute co-operation to ensure unhindered expansion. Delays such as the shaft-sinkers strike in 1935 outraged him. 'Wonthaggi's troubles [are] not political but industrial', he argued, caused 'by a section of the workers who need not be named'.[5] Nameless perhaps, but everyone knew who they were: they were the growing number of communist-led unionists whose influence dramatically increased after 1934. McKenzie determined to combat their growing social and industrial appeal, but in 1935 that confrontation remained in the future.

Solutions to the intractable problems of structural unemployment and industry consolidation provided a central theme for Federation campaigns for the next six years. In order to address these problems successfully, the Federation's national leadership – communist and left-wing ALP members alike – sought to reverse the process of regionalism that had developed after 1929, often with the tacit support of union officials and members. Wonthaggi is a case in point. As a smaller district, Victoria perhaps suffered more than other states from such an emphasis on regionalism. During the Northern Lockout of 1929–30, Victorian miners contributed fully to levies but two years later their own appeals for similar support fell on deaf ears. The 1934 strike demonstrated that Wonthaggi

---

5   *Sentinel*, 21 June 1935.

miners were 'regionalist' by default rather than by inclination. Minority Movement supporters at that time confidently expected Federation-wide support, and staked much of their reputation on such an outcome. One of the undeniable triumphs of the strike was Bill Orr's success in convincing all the other districts to strike a levy in support of Wonthaggi miners and to sustain it for five months. With others, Orr sought to use this newly won sentiment as a springboard for further national campaigns. Acutely gauging the temper of its members, the union's leadership developed a long-term strategy based on education, organisation and the involvement of the union's rank and file in support of specific demands. The Federation's paper, *Common Cause*, was revived as an independent eight-page weekly, released from its confinement as a supplement to the Lang-controlled *Labor Daily*. In Wonthaggi, the MM bulletin, *Miners' Voice*, suspended publication in July 1934 to reappear in November as *Union Voice*, a lively roneoed paper published fortnightly by the District and devoted to political comment, mining safety hints and Federation news, leavened with miners' own contributions and gossip from the pits. Publications such as *Union Voice* served a purpose in explaining Federation policy to members: how it had been determined, its intended aim and the place of all members in achieving it. This process of education and mobilisation culminated in 1937, and again in 1938, at national conventions that endorsed the strategy originally sketched out by Orr and Nelson in their three pamphlets on reconstruction. This campaign came to be known as the First and Second Rounds of the Log.

Behind such systematic preparation lay a deep-rooted urgency to implement overdue social reform while the opportunity existed:

> Shorter hours and higher wages are the immediate remedies for this national sore [unemployment]. This problem is facing the miners now and only the Log, successfully carried through, can prevent the mine-owners from getting a paralysing grip on the industry... If miners don't fight now when conditions are favourable they will, in the very near future, find that their means of livelihood are gone and they will be thrown on the labour market, not as unemployed but as unemployables.[6]

Summarised in *Coal Facts*, the Log represented a formidable set of claims: a 30-hour week underground and 35-hour week for surface workers, a 25 shilling per shift basic wage rate, an increase of 12.5 per cent in the

---

6   *Union Voice*, 27 May 1938.

contract rate, a Safety in Mines Bureau to research and regulate safer working conditions, a A£2 weekly pension for all retired mineworkers, and retirement at 60.

In the meantime, miners at Wonthaggi had continued to pursue industrial claims against a backdrop of deteriorating relations with the State Mine management. Since the depths of the Depression, the wages and conditions of these miners had been decided by a State Coal Mine Industrial Tribunal, and the Federation now used this tribunal to pursue local claims in support of national objectives. Wage rates proved an early target, and after 1934 previously declining wage rates had been successfully reversed. The tribunal granted marginal increases in 1934, and again in 1935. During 1936 the Federation mounted a full claim for the restoration of 1932 wage rates, including a 17 shilling and 6 pence minimum shift rate. Justice Winneke rejected the claim, relying on the argument that, unlike NSW mines, the State Mine lacked the capacity to pay such an increase. Miners rejected this 'dual rate' argument, claiming instead that ubiquitous features of the coal industry such as dangerous work and particular working conditions, overrode regional niceties. The unpopularity amongst miners of the tribunal also reflected its own difficult circumstances. Cumbersome in operation and often tardy in responding to claims and submissions, it couldn't cope with the pent-up flood of grievances released after July 1934. In addition to wage increases, union efforts concentrated on safety issues. Concern for safety reflected the universal belief that a miner's environment was uniformly hostile, a threatening world of 'darkness and dirt, poisonous and inflammable gases, falling stone, volumes of rushing water, a murky and dusty atmosphere and amid deeper mines, exhausting heat'.[7] The Federation also targeted mine mechanisation. At the State Mine, the miners' own sense of hearing and intuition often proved their best defence, allowing them vital seconds to escape falling rock or collapsing pit props. Miners argued that mechanical coal-cutters reduced these advantages. Similar concerns explain union opposition to any general speed-up in work underground, with miners caught between safer mining practices or fines for failing to 'make the min' (the amount of coal that a team needed to produce over a defined period to avoid incurring financial penalty). Reduced contract rates demanded higher shift production, while strictly enforced dirt fines (fines imposed when skips of coal contained too little coal and too

---

7   *Union Voice*, 4 September 1936.

much rock, or 'dirt') meant that corners would necessarily be cut to meet minimum efficiency levels. Such penalties inevitably led to a scramble on the next shift to make up arrears; those failing to do so risked dismissal as 'inefficient' miners. Twenty Shaft, in the Northern Area, achieved notoriety as a persistent trouble spot. Running disputes over safety issues led to the union enforcing the strictest possible interpretation of award conditions, a tactic met with summary fines and dismissals from an infuriated management.

Militant industrial action carried into 1937 as Victorian miners sought to emulate their counterparts in NSW, Queensland and Tasmania and regain pre-Depression wage parity. A protracted struggle to unionise smaller Gippsland mines met with significant success, often due to unorthodox yet highly effective tactics. Miners at the re-opened Kilcunda mine once again held Federation tickets. At Korumburra's Sunbeam mine, the local Branch president Wattie Doig organised and led a stay-down strike in September. The mine's manager had previously indicated his intention to ignore any new award and to defy him a team of miners barricaded themselves below ground. Participation in this stay-down strike by at least one Victorian District committee member as well as the Federation's national president, Charlie Nelson, indicates the degree to which the action reflected the union's industrial strategy. Only the need for urgent medical attention for one of their number finally brought them to the surface. Sunbeam miners received the wage increases management had intended to ignore.

Notable as an early Australian example of a stay-down strike or mine occupation, this Sunbeam strike also led to a dramatic pre-arranged response from Wonthaggi. When news of the strike broke in Wonthaggi, a stop-work meeting at the State Mine agreed to adjourn and march en masse to the aid of the Sunbeam strikers. Buses took the demonstrators to the outskirts of Korumburra, and from there 350 miners and their wives, led by the Wonthaggi Union Band, marched through the town to the Sunbeam pit head. At a public meeting, Williams and McVicars addressed a large sympathetic crowd. To consolidate local support, a Broad Committee and a Women's Auxiliary were formed on the Wonthaggi model. While the majority of mineworkers marched on Korumburra, two busloads led by District committeemen Bill Stirton and Bob Hamilton diverted to Jumbunna, where non-union labour had been hired to break a nine-week strike for union recognition and improved wages. Despite being menaced by armed men, Stirton's crew induced all non-unionists to stop work; 9 September 1937 proved to be a day of massive and

Figure 27. Striking miners emerging from the Sunbeam mine, Korumburra, after the 56-hour stay-down strike in 1937
The organiser of the mine occupation, Wattie Doig, wears a cap and stands in the centre at the rear of the group.
Wonthaggi Historical Society Collection

Figure 28. The Wonthaggi procession to the Sunbeam mine marches through Korumburra, led by the Wonthaggi Citizens Band*
*After the mid 1930s, the Wonthaggi Union Band was re-named the Wonthaggi Citizens Band.
Wonthaggi Historical Society Collection

Figure 29. Wonthaggi mineworkers and their families marching to the Sunbeam mine in support of the stay-down strikers, 1937
The 'Sunday best' clothes illustrate the exuberance of the occasion.
Wonthaggi Historical Society Collection

popularly supported action, particularly evocative of the style, strategy and success of the Federation's new militancy.[8]

Six months previously, a local *Common Cause* correspondent (possibly Idris Williams) had returned to a well-rehearsed theme: the decline in safety standards and the increase in accidents at the State Mine. The article appeared partly in response to a management decision to suspend a young clipper (miner's assistant), Pollock, preparatory to his dismissal for being accident-prone. In union vocabulary, 'accident-prone' meant being concerned with safety issues and acting on such concerns. On Monday 15 February, after a week of deadlocked negotiations over Pollock's dismissal, a large public meeting organised by the Federation assembled at the Union Theatre to agitate for increased safety. A little before 10.30 am, with the Wonthaggi Citizens Band still playing to summon a crowd, a mine bus drew up outside the theatre. From its step, George Lees, leader of the mine's rescue team, appealed for volunteers. Fifteen minutes earlier, prophecy had become fact. Twenty Shaft had erupted:

> Roaring like a tornado along the drive and splitting to race in two opposite directions giving none of the men a chance, the ignited gas wrought great havoc in the mine workings… thirteen men were trapped behind thousands of tons of fallen rock, loose coal and twelve inch pit props which had been ripped from position and hurled along the drives with incredible force.[9]

Amid highly emotional scenes, streets emptied and businesses closed as hundreds of locals milled around the Twenty Shaft pit head while rescue teams, still at great risk, explored the extent of the explosion. As general manager, McLeish assumed responsibility for the first exploratory survey, and

> with Thomas Johnston, the underground manager, Ron Speirs, a brother of one of the entombed men and Ted Fairless, an experienced miner… he entered the cage [which] disappeared into the blackness of the shaft. Gongs rung by a miner… signalled the descent of the cage toward the bottom, and, with each gong, the strain of waiting became more exacting. Miners assembled around the pit head watched the moving hawser without speaking. Their faces told their anxiety.

---

8 For contemporary accounts see *Common Cause*, 18 September 1937; *Union Voice*, September 1937.
9 See, especially, the *Powlett Express*, 19 February 1937, for eyewitness accounts of the disaster; see also accounts in the *Sentinel*, *The Age* and the *Sun News-Pictorial*.

Then, after minutes which seemed like hours, three gongs sounded for the cage to be stopped. The signal meant, 'Stop, danger'. As the gongs sounded every spectator shouted, 'Hold her'. The hawser stopped. The cage had sunk to within 16 feet of the bottom of the shaft but could go no further. To the men in the cage the horror of the explosion became poignantly apparent. Even the skids on which the cage ran were blown away. Before them in the tunnels piles of coal and splintered woodwork, which had been supports, revealed the terrible force of the explosion.[10]

Not until mid afternoon, when the first bodies were brought to the surface, did Wonthaggi comprehend the full extent of this disaster: 13 men – two overmen (shift managers), six deputies (under managers), three pumpers and two ropemen – had died. No-one below ground at the moment of explosion survived, which is not surprising given the extent of the blast:

It is believed by officials of the mine that the force of the explosion must have travelled nearly a mile from where it took place before it reached the pit-head but it hurled the heavy cage 30 feet up to the poppet-head from the surface, damaging the skids on which it ran and buckling the thick, steel hawser on which it was suspended... A mile away the concussion broke every article of crockery in the home of Mr Keith Hollole, and shattered the windows of his kitchen and dining room. Every shutter in the telephone exchange at Dalyston, two miles away, was dislodged by the explosion. The cloud of coal dust shot 100 feet into the air from the pit-head, and drifted across the countryside without breaking.[11]

While immediate differences were put aside in the interests of rescue and recovery, in the longer term this disaster heightened antagonism within the community and effectively undermined the limited common ground between miners and management that had been laboriously rebuilt since 1934. Both parties adopted rigid and unyielding positions, while also disclaiming responsibility for disruption or strike action at the mine. Convicted in the court of public opinion, the mine's management – McLeish in particular – suffered further when the official inquiry into the disaster reported in early April that

we find that death was accidentally caused... there is no evidence to show where the explosion started or by whom the naked light was

---

10 [Melbourne] *Herald*, 15 February 1937.
11 [Melbourne] *Sun News-Pictorial*, 18 February 1937.

Figure 30. Buckled rails and smashed skips: evidence of the force of the explosion at Twenty Shaft, February 1937
Wonthaggi Historical Society Collection

Figure 31. The State Mine's general manager, John McLeish, inspects damage at the bottom of Twenty Shaft in the aftermath of the explosion, February 1937
Wonthaggi Historical Society Collection

Figure 32. Stretchers being taken below by a rescue party to recover the bodies of victims a week after the explosion, February 1937
Wonthaggi Historical Society Collection

Figure 33. Press coverage of the Twenty Shaft disaster: rescue teams emerge with bad news
Wonthaggi Historical Society Collection

Figure 34. Members of the State Coal Mine Rescue Brigade, 1937
Wonthaggi Historical Society Collection

Figure 35. Aerial view of Twenty Shaft only hours after the explosion
Miners demanding a place in the rescue teams besiege the pit head, while on the Melbourne road outside hundreds gather to hear news of those caught below.
Wonthaggi Historical Society Collection

Figure 36. An image of chaos in the Twenty Shaft workings
Members of a rescue team inspect a skip torn apart by the cyclonic force of the explosion, February 1937.
Wonthaggi Historical Society Collection

Figure 37. Members of the State Mine rescue team after a shift underground following the Twenty Shaft explosion, February 1937
Wonthaggi Historical Society Collection

Figure 38. Refreshments for rescue team members following a four-hour shift underground seeking survivors of the Twenty Shaft explosion
The shift had just finished at midnight and the members went below again four hours later.
Wonthaggi Historical Society Collection

Figure 39. Members of the inquest into the Twenty Shaft mine explosion confer with local police, February/March 1937
Wonthaggi Historical Society Collection

Figure 40. Refreshments for rescue workers following a shift underground, February 1937
Mrs Haughton stands at the left organising food and cups of tea.
Wonthggi Historical Society Collection

carried. We also find negligence on the part of the management in not having the mine inspected on that day in the same manner as on the ordinary working day shift.[12]

A later finding by the subsequent Royal Commission exonerated mine management on this issue, but this did little to change local sentiment.

Recriminations delayed any resumption of work until May, as the Federation stood firm on the principles of mine safety and defence of established mining practice. In pursuit of these principles, it refused any return to work without a guarantee that no retrenchments would occur and that separate cavils (or ballots for work sites) would not be imposed for individual areas at the expense of a general cavil for the entire mine. Miners feared that McLeish might seek to use the disaster for industrial advantage. They could point to the agreement of February to resume with an additional miner at each workplace, rescinded on the morning work was scheduled to resume. Or the unilateral demand, after a cavil for workplaces had finally been held in late April, that 100 men transfer to Twenty Shaft. By then, even the local press had asked: 'is the management endeavouring to force the men into a strike and thus… justify themselves under a pretence?' 'Certainly', miners replied, burning McLeish's effigy before an angry crowd of 600 at a May Day rally in Graham Street. [13]

Throughout February a series of funerals, each led by the Wonthaggi Citizens Band and accompanied by hundreds of miners, wound their way from Church Hill, south of the town centre, down Hagelthorn Street then south along Cameron Street to the pine-fringed cemetery on Wonthaggi's southern outskirts. Wonthaggi mourned, and grief in turn gave way to bitterness and cynicism. The community questioned not only the competence of mine management but also the irrationality of mining life itself. In the Women's Auxiliary speakers' notes from April, these themes stood out:

THIRTEEN MEN KILLED

Explain the position of women at that time owing to idle time.

Xmas holidays without pay and the scrimping resulting.

---

12 For considerations and investigations in the aftermath of the disaster, see the text of the jury of inquiry verdict handed down on 1 April 1937 and published in the *Sentinel*, 9 April 1937 and 1937. 'Interim report of the Royal Commission on certain matters relating to the State Coal Mine', Wonthaggi. Melbourne: Victorian Government.
13 See the *Sentinel*, 30 April 1937; *Union Voice*, June 1937.

THE POSITION OF WOMEN AND PAYING BILLS AT AND DURING THE TIME OF EXPLOSION.

Dire poverty.

Not able to make ends meet resulting in kiddies going to school undernourished and sometimes bootless.

Quote cases of sores being prevalent among school children.

Do our children get proper nourishment? If not, what is the cause? The attitude of the management.

IS IT FAIR THAT OUR HUSBANDS, SONS AND BROTHERS SHOULD BE FORCED TO WORK UNDER THE MANAGEMENT WHICH HAS BEEN FOUND NEGLIGENT BY A JURY OF NINE?[14]

The Federation withdrew from the Royal Commission into the disaster on the second day of hearing in protest of its restrictive terms of reference – claiming it to be little more than an expensive whitewash. But it kept a close watch on safety issues, agreeing unofficially to continue to participate in the Royal Commission's deliberations on new mining regulations for the State Mine and, as a consequence, successfully arguing for the adoption of many of its own recommendations on mine safety.

A wave of public sympathy meant that a relief fund for families of the victims was able to raise more than A£29,000 within a week, but after five years of having to rely on self-support and local community assistance for survival, such a show of press and public sympathy for the miners of 'Red' Wonthaggi seemed unusual enough to be the subject of street chatter, while *Union Voice* went so far as to query whether such sympathy would be forthcoming next time miners went on strike. Even so, public pressure combined with economic necessity to force a compromise settlement. McLeish would not be dismissed and work resumed on 17 May. However, with the exception of the Western Area, the entire State Mine had been gazetted as a safety mine – naked flames for any purpose were forbidden and replaced by safety lamps, shift rates replaced contract rates, and although McLeish lamented that production had fallen 40 per cent as a consequence, miners celebrated that they had 'passed through a

---

14 'ACSEF Women's Speakers notes', n.d. (April 1937). Rankine Collection. University of Melbourne Archives.

short period that has actually been the best for conditions that this mine has known'.¹⁵

On a national level, Wonthaggi miners emerged as the vanguard of the Federation's safety campaign. This campaign sought to convert 'public alarm and indignation' at the disaster into support for a wider log of claims. The Federation's national focus emphasised the threat of mechanisation, but on the Powlett field inconvenient geology rather than union opposition frustrated increased mechanisation. Management pursued the possibility of increasing mechanisation, but found itself frustrated at every turn. Mechanical coal cutters had been trialled in the Northern and Western areas during 1937, testing management's estimate that one two-man machine could produce 50 per cent more coal than 20 miners working manually. Only in the Western Area did the machines prove even marginally economical, and in Wonthaggi only two machines could be installed. Geology ensured that the State Mine would remain the preserve of contract mining.

Geology aside, Wonthaggi's miners participated fully in the meticulous planning for the upcoming Log Campaign. The Federation sought to enlist the widest possible community support, and strike preparations had been completed well in advance. Throughout the first months of 1938, newspapers in Wonthaggi, Korumburra and Melbourne carried a series of articles written by Idris Williams summarising the union's case:

> We have 21,000 mineworkers demanding these reforms, and we have learned how to struggle intelligently, and by organising the general public and workers of this country… The case placed forward by Mineworkers is based on facts. The mining communities of this country must not be allowed to become derelict… Therefore, for a better community with better opportunities, let us go forward in unity for a better social standard of culture and a better life for all dependent upon the mining industry.¹⁶

Anticipating possible strike action, the Federation secured A£250,000 in credit. A commitment of A£10,000 from the Wonthaggi Co-operative Society underwrote Victoria's share of A£15,000, with the balance guaranteed by union auxiliaries and sympathetic businesses in and around the town. Additional moral and financial support came from Victorian unions and Trades and Labour councils.

---

15   *Union Voice*, September 1937.
16   Idris Williams, untitled article later reprinted in the *Sentinel, Powlett Express, Korumburra Times* and *Common Cause*.

The Federation's August 1938 National Convention endorsed the final version of the Log of Claims, which was subsequently served on the Commonwealth Government, state government and individual proprietors. An attempt by the Railway Commissioners to have the State Mine exempted from Federation claims failed and, in turn, the commissioners rejected the union's case. The subsequent six-week strike climaxed three years of preparation and planning. The Federation recognised the role of mining communities as a strong factor in its favour in any dispute, capable of mobilising support far beyond the union itself, while also serving as a means of taking some of the tension out of the increasingly fraught relations between communists and ALP activists. This campaign for community support gained momentum throughout 1938 and reached its peak in late July in a mass public rally through Wonthaggi's streets. Fourteen months previously a mob of 600 had publicly burned McLeish's effigy. Now, on 31 July, an exuberant crowd of more than 2000, confident in its capacity to win, paraded under the slogan: 'Action Wins, a Step Nearer Socialism', in the biggest demonstration of industrial support since 1934. Already endorsed by borough councillors and local ministers – some of whom had been smuggled underground to satisfy themselves of the justice of the miners' demands – the campaign culminated that day in a procession led by the Wonthaggi Citizens Band, followed by Wonthaggi, Kilcunda and Korumburra miners, the Miners Women's Auxiliary, the Timber Workers Union, the Engine Drivers and other unions, the local unemployed, contingents from Friendly Societies, the ALP, the Communist Party and other sympathisers in a demonstration of popular support few other communities could match.

In the meantime, the longer term effects of the 1937 disaster continued to drive mine reform. Although the second report of the Royal Commission into the disaster, tabled in late 1937, carefully noted that most of the reforms to the Victorian mining code proposed by the Federation 'have been adopted', unionists begged to differ. Basing its recommendations on current NSW statutes, the Commission recommended a range of more stringent safety levels and minimum standards of mining practice, including improved ventilation, better inspections for dust and gas and a strict code for shot-firing in both normal and safety pits. But eight months later, as the campaign for the Log reached its climax, union propaganda asked rhetorically: 'A Royal Commission… inquired into the disastrous explosion in Twenty Shaft, and made certain recommendations for the safe working of coal mines in Victoria. What has become of them?'

The six-week national strike in support of the Log ended when four of the Federation's primary claims – those relating to hours of work, wage rates, paid holidays and weekly pays – were referred to arbitration, with a Royal Commission established to consider mine safety and health issues in greater detail. All parties to the dispute recognised that the impending determination would pass definitive judgement on the state of the industry, and the future of union claims to a voice in any industrial restructuring. When handed down in June 1939, Justice Drake-Brockman's interim award dashed the Railway Commissioners' hope that the State Coal Mine would be considered as a special case. Drake-Brockman turned the commissioners' own case for the existence of the State Mine back on them, arguing that

> any losses [from the State Mine's operations] represent the premium that Victoria is prepared to pay as an insurance against uninterrupted supplies… Whether or not that premium is justified is a matter for the consideration of the Victorian Government and not of this court.[17]

He did not stop there. In a vindication of the Federation's campaign against 'exceptionalism' in Victoria and other smaller districts, Drake-Brockman accepted the argument that the coal industry, 'because it is a national industry, should be controlled on a national basis and by an Australian tribunal rather than by un-coordinated awards of State Tribunals'. In a win for the Federation, Drake-Brockman granted miners 10 days paid leave annually, conditional on the regularity of their work. Wages would be increased, but not paid weekly. He rejected the union's ambit claim for a 30-hour week for underground workers, ruling instead for an eight-hour day and five-day week. Hours would remain unaltered for surface employees, much to their disgust. Although debate around the supplementary issues continued well into the war years, by early 1940 it had become apparent that Australian coalminers had won their most significant economic gains for a quarter of a century, regaining the industrial ground lost during the Depression.

But for the State Coal Mine, and its workforce, Drake-Brockman's decisions proved a double-edged sword. Lower wage rates and lesser working conditions had been essential to the economic case that mine management and state governments had argued for the past decade. Such a strategy was not necessarily the only way to ensure that the State Mine's potential, however limited, would continue to be tapped. The key to issues such as future capital investment and developmental work lay in the 'premium', or

---

17　Australian Coal and Shale Employees Federation [Miners Federation]. 1939. Extract from the *Interim Award*. Wonthaggi: ACSEF Victorian District: 21.

the value of the State Mine to the state that Drake-Brockman had referred to in his judgement. In this respect, Drake-Brockman's determination posed a long-term threat to the State Mine. For Wonthaggi, the question now became not whether lower wages and lesser conditions could keep the State Mine open, but whether the asset was needed any longer.

The political dynamics of Wonthaggi had changed fundamentally in 1934. Success in leading the five-month strike lent militant miners, and the communists among them, industrial credibility and a measure of political respect that had not existed previously. In Victoria, Wonthaggi became one of the 'hot spots' where the confrontation between communist activists and the ALP would be played out in the late 1930s as well as during the war. Communist organisation in Wonthaggi appeared relatively late when compared to other mining districts. It was not until 1933 that either a Communist Party branch or an MM unit was successfully established among State Mine employees. The reasons for this were varied, but the radicalising effect of the Depression and disenchantment with the performance of the ALP, particularly when in government, proved common themes. In the words of one local miner who joined the Communist Party in 1933: 'I left a Labor Party meeting at which the [Premier's Plan] policy was endorsed 7 votes to 5 and never returned to any other meeting…'[18]

Yet it would be misleading to imply that all those miners, whether British immigrants or not, who so enthusiastically endorsed the Federation's new strategy and who joined the MM became members of the Communist Party. The Communist Party could never recruit more than a small minority of miners, at the State Mine or elsewhere. Jim Chambers, for example, was never considered a particularly expert miner. He won respect, instead, as a socialist, an adjudicator to whom crib-time arguments would be referred. In Scotland, the Chamberses had been active members of the Broxburn Independent Labour Party (ILP) branch, and their home had sheltered John McLean and other socialist propagandists during Scottish speaking tours. Chambers never joined the Communist Party. Like so many other British mineworkers in Wonthaggi, he accepted the legitimacy of communist industrial objectives and their strategy for the coal industry while rejecting party membership. Chambers's own disagreement was tactical, for he remained suspicious of what he saw as the pre-emptive influence of working within the Communist Party, believing that it ultimately served to stifle discussion and narrow political options. But such reservations proved no obstacle to industrial co-operation, even for many miners who remained in the ALP. Following 1934, a communist-led union

---

18 Bill Stirton, 'Depression in Victoria', mss notes provided to the author, 1974.

Executive encouraged a militant, community-based strategy, supported by a coalition of communist, ALP and non-party mineworkers, cemented by their common acceptance of trade unions, particularly their own, as instruments of social as well as industrial reform.

The drift into communism would never be uniform: some left the ALP but joined no other party; others adopted a militant industrial position with no party affiliation; and still others voiced their disenchantment while remaining members of the ALP. Whatever the personal choices of miners in Wonthaggi, it is clear that the events of 1934 allowed communists and their sympathisers to win not only a mass base among the State Mine workforce but also a majority of the Victorian District's Executive and committee positions. But communist control had its limits. Relying as it did on effective collaboration with some ALP members and the bulk of the union's membership who had no formal party affiliation, communists could not afford to lose sight of the fact that for the majority of miners, their commitment to the union, even as a vehicle for political activity, remained stronger than any party affiliation. Theirs had to be a 'practical communism', reflecting the industrial priorities of the union's membership. The delicacy of their situation can be seen in the way Idris Williams and other communists consolidated their position after 1934. On a number of issues that had a rhetorical as well as a practical value, a communist preference can be seen. Examples include the decision to affiliate with the MM in September 1934, to join the Victorian Council Against War and Fascism (VCAWF) – an organisation proscribed by the ALP – in January 1936, and to nominate the Victorian secretary of the CPA, J D Blake, as the District's delegate to the Melbourne Trades Hall Council (THC) in the same month. But in the wider industrial sphere – in campaigning during the First and Second Rounds of the Log, for example – communist strategy consciously sought to avoid ideological confrontation within the union. In mobilising support, the communist leadership deliberately appealed to citizens rather than workers, to communities rather than classes.

While collaboration across party lines remained a continuing priority in industrial affairs, no equivalent accommodation could, or would, be reached between the two political parties. No matter what overtures came from the CPA, Labor parliamentarians and officials remained staunchly anti-communist, as scathing of fellow travellers in their own ranks as of communists themselves. The efforts of communists to expand their political and community influence after 1934 brought them into direct conflict with entrenched ALP interests. By the mid 1930s the ALP, too, had begun to recover its balance, both politically and industrially, and growing communist influence would not go unchallenged.

In 1936 Wonthaggi was still far removed from the thriving confident community it had been in 1929. Although emerging triumphant from a major strike, the town still faced a struggle for survival. Declining coal production remained a constant source of concern. By 1939, State Mine employment had only barely recovered to the level following the mass sackings in 1932, and even this number had been inflated by an extra 200 miners employed in 1938 to stimulate lagging developmental work. Wonthaggi had never really recovered from the 1932 retrenchments. The town's population stagnated throughout this decade, with the local press carrying regular reports of families leaving the district. The state of the Borough's finances reflected the debilitating impact of the Depression on local government. Throughout 1934 the Council consistently devoted at least 30 per cent of its diminished revenue to relief work, although by late 1935, with the completion of a local road-surfacing program, relief opportunities dried up. The pressure on borough finances can be viewed in a report from early 1935 revealing that in the preceding year the community had paid A£3591 in rates, while a further A£3128 remained unpaid.

With limited local opportunities, Wonthaggi's unemployed found themselves required to travel up to 400 miles, at times as far as Sunraysia, to help construct underground irrigation tunnels. Those failing to take up such work lost their eligibility for sustenance relief. Too often the unemployed found regular relief work elusive, conditions of employment unsatisfactory and standards of safety and hygiene non-existent. Influenced by the success of the miners' strike, unemployed locals struck in March 1935, claiming an increase in the dole, an adequate clothing allowance, and the provision of firewood for relief work campsites. The strike itself met with mixed success, winning a marginal increase in the dole but failing to extract any guarantee of regular relief work.[19] But significantly, their actions now took the form of a strike rather than a delegation seeking assistance. In Wonthaggi at least, the unemployed still saw themselves as an organised element of the working class.

The prominence of communists in unemployed strike action led to the first acute political tensions in Wonthaggi since the 1934 miners' strike. Since that strike, Bill McKenzie had accepted the reality of working with Williams and other communists in their capacity as union officials, but he

---

19   *Sentinel*, 29 March 1935.

could not tolerate their influence with the unemployed. He rejected UWU appeals for his support, for 'he would speak at no gathering in support of the sustenance workers strike. That very strike had nearly ruined the chance of getting the new rates now operating'.[20] McKenzie had long warned of the political consequences of unemployment in Australia, and he now saw those concerns being realised. Only during the course of the 1934 strike did he recognise the scale of this threat among the ranks of Wonthaggi's employed as well, and had unsuccessfully worked to secure the return of the 'constitutionalist' ticket at the June 1934 elections. He now had to watch as the Federation, under a leadership opposed to his own views, reclaimed its civic influence. The place of public meetings in Wonthaggi's political life had long been recognised, as had the relatively subordinate role occupied by the Federation in such events. Generally, these meetings had been staged under the auspices of a 'neutral' institution – the mayor, Council itself, a Progress Association or one of the Public Sale of Coal organisations that had organised and re-organised over the years. Elements of this practice survived, but the Federation now assumed a much more active role. A single Executive member recruited to sit on a stage under another organisation's banner no longer sufficed – many public meetings would now be called by the union itself, with its members attending to debate and support union-sponsored initiatives.

When, in March 1935, the Railway Commissioners responded to renewed industrial action by threatening to abandon all further developmental work at the mine, the Federation responded in classic style, organising the Wonthaggi Advancement and Defence League. The emphasis on advancement harked back to earlier agitation for public coal sales: indeed, one League objective committed the organisation to 'the continued development of the mine'. Yet the next objective caused some to hesitate: 'the defence of the rights of the people'. Defence of what, against whom? McKenzie could recall earlier defence leagues, the communist-led Workers Defence leagues of three or four years earlier which numbered the ALP among its principal enemies. Nevertheless, he accepted a position on the League's general committee, along with members from the Federation, the Traders' Association, the Unemployed Workers, the ALP and the CPA.[21]

---

20   *Sentinel*, 12 April 1935.
21   For the development of the Wonthaggi Advancement and Defence League, see the *Sentinel* and *Union Voice* during the period March to June 1935.

Taking up the cause of the underpaid shaft sinkers at the State Mine, in defiance of a State Coal Tribunal determination, set the tone of the League. The influence of the Federation, and by extension the Popular Front strategy of the union's leadership, soon became evident. The League served as a catalyst, organising hitherto disparate, often antagonistic community groups in support of well-crafted union objectives, most notably the emerging Federation campaign against indiscriminate mine mechanisation. The breadth of the League's interests troubled McKenzie and many borough councillors. They had hoped to restrict the League to industrial matters; now it appeared that the Federation, and its communist leaders, had acquired a potent political instrument. At this stage, McKenzie moved to redirect the League's activities. It had been

> set up by the Miners' Union to put weight on to settle trouble at the mine. It had done that and now appeared to be going further than originally intended... Now it was doing, or trying to do things which were already being done by the Progress Association, the Traders' Association and the council... They are now setting out to do things and take actions in matters which they had no authority to do.[22]

McKenzie's attack signalled a trial of strength between the ALP and communist-led union militants, a resumption of the bitter campaigns waged within the Federation's Wonthaggi Branch a year earlier. As a consequence, the League disintegrated, its functions progressively incorporated into the mine safety agitation of 1937 and the Second Round Log campaign during 1938. But the political objectives of the League promised to be more enduring. McKenzie did not, indeed could not, object to union defence of industrial conditions – he encouraged this – nor could he deny the Federation's interest in mine development and markets for State Mine coal when he himself had been instrumental in promoting just such a campaign a decade previously. But he remained implacably opposed to the Federation's usurping of what he understood to be the legitimate functions of other community organisations. His stance, obviously enough, had a sharp political edge: he 'did not approve of Communists having a voice in the League, personally he was not happy sitting with communists'. Although the League dissolved following McKenzie's defection, he was not fighting a single organisation but a broader political strategy. Despite his efforts, the Federation had now

22 *Sentinel*, 21 June 1935.

embraced a new political activism, one driven, as he clearly recognised, by the Communist Party.

For more than a decade, Wonthaggi's politics revolved around the communist-supported Federation Branch and the local ALP Branch that gained its authority from McKenzie's parliamentary status. Through the Federation, communists worked to involve disparate organisations, irrespective of political allegiance, in pursuit of community and union objectives. McKenzie, in turn, decried such overtures, arguing instead that: 'the Labor Movement was wide enough to do all the Communists sought... The only thing the other party did is to belittle the actions of Labor... [in] an effort to destroy the Labor Party'. By virtue of his parliamentary status McKenzie dominated the local ALP Branch, a branch now reduced to a shadow of its former strength. By mid 1936 its membership stood at less than 70, while three years later the Federation opposed a qualification of two years membership to qualify as an ALP conference delegate because that would 'confine the selection of delegates to less than twenty members of this union'.[23] Unlike many of his Labor contemporaries, McKenzie had survived the political setbacks that beset the ALP during the Depression. By 1935 he could again rely on an absolute majority at any election, despite the regular presence of a communist candidate. Although communists received sufficient votes to have their deposit returned when they stood for Wonthaggi, they never seriously challenged McKenzie. What proved more disturbing was the electorate's slow drift from what was essentially an urban electorate to one in which rural subdivisions represented the majority. McKenzie's grip on non-Wonthaggi booths also loosened; from parity with the Country Party at Korumburra in 1935, by 1947 he would be outpolled there by two to one.

Unable to challenge McKenzie at the polls, the CPA sought instead to denigrate the ALP, suggesting in 1936 that 'the fact that the workers of Wonthaggi did not turn up [to an ALP rally]... shows that the Labor Party has lost its grip on the working-class'.[24] This was pure rhetoric, for the ALP's resilience lay not in its actual membership but in its core electoral support and parliamentary representation. The Federation well knew the importance of maintaining McKenzie's active co-operation in the political and industrial affairs of the State Mine, and he seldom let them forget this: 'while every member of the miners' union is entitled to attend branch meetings as

---

23   *Union Voice*, 5 May 1939.
24   *Sentinel*, 3 July 1936.

affiliated members, few thought it worthwhile… But so soon as trouble arose it was the Labor Party which was called upon to settle it'. In fact, much of the tension in Federation–ALP relations during this period can be traced to the Federation's active interest in this party that had 'lost its grip'. The Federation urged its own policies at ALP conferences, and ultimately cut its formal links with the VCAWF in order to retain ALP affiliation. Similarly, at the 1938 ALP State Conference, Federation delegate credentials were successfully challenged. This could hardly have come as a surprise, given that at least one of the two delegates was a recognised communist supporter, if not a secret party member. Such a stormy relationship marks the ambivalence of the popular front politics practised by the Communist Party. For the bulk of the ALP, the CPA remained anathema. For communists, matters were more complicated. With co-existence hardly possible, they faced an alternative of extremes. If they could not woo the ALP into a political alliance, then they had to defeat it electorally. Ultimately, the Communist Party failed to achieve either. Consequently, its tactics could veer erratically. In local community involvement, conciliation often proved to be the preferred tactic, and through this method, rather than any wider political activity, the strategy of closer co-operation came closest to success.

In the years following the 1934 strike, communist organisation in Wonthaggi matured. The fragile cell established in 1933 developed into a significant party branch, dominated by State Mine employees. The actual membership is difficult to estimate, although an indication of support can be seen in the crowd of 400 that attended a Communist Party 'Defeat the Lyons Government' dance in July 1937. A kaleidoscope of subsidiary organisations appeared in Wonthaggi: the VCAWF in 1935, the Workers' Sports Federation the same year, the League of Young Democrats two years later and a Left Book Club discussion group in late 1938. Of these, two organisations stand out as particularly significant: the Wonthaggi Miners Women's Auxiliary (organised in 1935 by the Federation, not the Communist Party) and the VCAWF.

The Women's Auxiliary organised female members of miners' families, building on the earlier efforts of the innocuously named Wonthaggi Social Club – originally formed by British miners and their families in the 1920s – and later from the leading role women took in a number of the Broad Committee's operations during the 1934 strike, including touring with the Propaganda subcommittee's speaking teams. In fact, during the latter stages of the strike women had formed a parallel Broad Committee of their own, concentrating on delivering food and other support to women and children in particular need. As Agnes Chambers, the Auxiliary's first president, argued, 'the Government threatens to close the mine permanently. This will mean the breaking up of our homes and the ruination of the lives of our children. Our women have assisted to organise a well conducted system of relief'. Born in Broxburn, Lanarkshire, in 1883, Agnes Chambers is remembered as a fluent speaker and popular leader of the Auxiliary. A member of the ILP and the Women's Co-operative Guild in Scotland, Chambers shared the political schooling and outlook of her immigrant generation. Like many others of her generation she read widely, and cited Beatrice Webb, H G Wells and Edward Bellamy as youthful influences. Migrating to Australia in 1922 with her miner husband, Jim Chambers, she became active initially in the local co-op women's group in 1924 and later the State School Mothers' Club (serving as president in 1927) before assuming a prominent role in union publicity and propaganda during the 1934 strike.

The Auxiliary became the most significant and influential community organisation established in Wonthaggi after 1934. It acted as an anti-war and anti-fascist group, placing great emphasis on international causes, evident in its championing of a boycott of Japanese goods after 1936, and strong support for Republican Spain. From its earliest years, communist organisation in Wonthaggi was marked by a high degree of women's involvement. Instances abound of both husband and wife, or parents and children being involved in common political activities, a circumstance from which the Auxiliary and the Federation at large gained considerable advantage. The Auxiliary also recruited other local women as well as farmers' wives from the surrounding district. Although it did not exert a direct influence over Federation industrial strategy, the Auxiliary never sought to hide either its political character or its willingness to engage in industrial issues. It believed itself to be proof of 'the inestimable value of a women's organisation when affiliated to any trade union... remember, the [Auxiliary] is our union'.[25] Acting on such sentiments, the Auxiliary financially supported striking miners at Korumburra and Jumbunna and assisted strikers at the Lysaght steel mill in New South Wales. It also provided aid to local needy families and assisted the 1939 bushfire relief appeal. Auxiliary members resumed their propaganda role in 1949, contributing eight members to the union's campaign for public support in Melbourne and Geelong.

Given its social composition, the Auxiliary also emphasised the primary importance of community issues – its campaigns for a maternity wing at the

---

25 *Union Voice*, 10 March 1939.

Wonthaggi hospital, for mine safety following the 1937 explosion, and for a kindergarten for the town are just three examples of their activity in this regard. In many instances, the Auxiliary acted on its own behalf, initiating and sustaining its own campaigns, however complementary they may have been to broader Federation objectives. Its organisational methods proved familiar and successful: a socialist group using community organisation to press a social agenda. It placed great emphasis on education and information, and public meetings of up to 500 women became a regular prelude to any Auxiliary campaign, a reflection of the common cause the Auxiliary had established with other women's groups in the town. As one of the first auxiliaries to be established in any industry, the Wonthaggi Auxiliary became a prototype for similar organisations in other coalfield communities and those sponsored by other militant unions, most notably the Waterside Workers, the Railway Workers and the Seamen.[26]

For its part, the VCAWF tapped a latent anti-war sentiment shared by many miners who had served with the Australian or British forces during the First World War and who now remembered those years with a mixture of cynicism and disquiet. Such views did not remain the province of anti-war groups; even the local branch of the RSL, when considering compulsory military training in August 1936, rejected the proposal by 47 votes to 4. But the VCAWF offered the most vigorous criticism, such as its response to the same issue of military training. As the local secretary wrote:

> I, also, possess a 'number of ribbons for service' which I treasure as a constant reminder of the fool I was in 1914. A list [for recruits] is awaiting signatures in the Town Hall. May the youth of Wonthaggi avoid the Town Hall like the plague, and by so doing, show the Lyons Government that we want peace, bread and a happy life, not war…[27]

Members of the Women's Auxiliary also worked to establish a district anti-war committee, and some members were also active in the Local Affairs Committee established by the CPA in September 1937. This

---

26 For further information on the work of the Women's Auxiliary and Agnes Chambers, see Joyce Stevens. 1987. *Taking the Revolution Home: Work Among Women in the Communist Party of Australia, 1920–1945*. Melbourne: Sybylla Co-operative Press; Peter Cochrane. 1973. 'The Wonthaggi coal strike, 1934'. B.A. Honours thesis. Melbourne: La Trobe University; Andrew Reeves. 1986. '"Damned Scotsmen": British migrants and the Australian Coal Industry, 1919–49'. *Common Cause: Essays in Australian and New Zealand Labour History*. Sydney: Allen and Unwin; as well as subject files on the Auxiliary and Agnes Chambers held by the Wonthaggi Historical Society.
27 G A Allen, *Sentinel*, 7 August 1936.

committee's formation coincided with a communist move to contest borough elections. In August the Communist Party nominated Bill Stirton, formerly a Federation vice-president and now a union check-inspector at the State Mine, to challenge Councillor Easton, the leading conservative councillor. Stirton had to withdraw his nomination upon discovering that he lacked a necessary property qualification to nominate, but the lesson of his premature candidacy stood, and bore fruit two years later.

These years preceding the renewal of world war reflect a political stalemate in Wonthaggi. McKenzie's political position resisted all challenges, but, in turn, he could no longer deflect the Federation from its industrial or political strategies. As ever, strike action remained a bone of contention. Although publicly supporting Federation industrial claims between 1937 and 1940, he remained an outspoken opponent of Federation tactics, becoming particularly angry at the Federation's decision in 1938 to ban developmental work in the Kirrak Basin during the six-week strike. Antipathy toward such Federation decisions provoked one final prewar confrontation in the town. As early as 1936, Labor commentators had lamented the relative under-development of unionism beyond the limits of the State Mine. Union organisers working among the Gippsland timber cutters supplying the State Mine regularly spoke of the dozens of workers who remained non-unionists. Similarly, in the heart of 'Red' Wonthaggi, shop assistants and hotel employees needed organising. Many in the ALP believed the answer to be a local trades and labour council, which by organising previously non-unionised workers could also restore its industrial influence in Wonthaggi. A successful ALP-inspired timber workers' strike lent the idea momentum. The formation of the Wonthaggi Trades and Labour Council (TLC) in January 1939 represented a proud moment for local ALP members. At its inaugural meeting, attended by 'Dinny' Lovegrove, the ALP state secretary and J V Stout, secretary of the Melbourne Trades Hall Council, the local council pledged itself to industrial stability and solidarity in Wonthaggi. The Miners Federation headed the list of affiliates, joined by the FEDFA, AEU, Carpenters and Joiners, Liquor Trades, Shop Assistants, Municipal Workers, ARU, AFULE, Firemen, and Hospital Employees. If McKenzie or other ALP members believed that the TLC could harness the Federation, and with it communist influence, they were soon to be disappointed. The Federation responded enthusiastically to the TLC's establishment, with Idris Williams, District President and avowed communist, unanimously elected inaugural president. With the active co-operation of other unions covering the State

Mine's workforce, such as the FEDFA and the Engineers, Federation delegates worked assiduously to involve the TLC in the latter stages of the Log campaign. At the moment of the TLC's creation, Williams clearly defined the limits of its influence. On industrial matters, the Federation would not be dictated to by the TLC. McKenzie remained silent, for such control was indeed his aim.

As previously described, Drake-Brockman's interim award rejected a claim for reduced hours by surface workers, on the grounds that they did not meet the definition of dangerous work. In response, surface workers downed tools. McKenzie sought to use the constitution of the TLC to bring such a 'frivolous' strike to a speedy end, claiming that the Federation had to accept the TLC as the 'governing and deciding body in the town on industrial matters'.[28] His challenge met with absolute refusal. In Federation eyes the TLC had been established to improve industrial effectiveness, not hinder it. The surface workers' strike continued to a resolution by negotiation, but the possibility of the TLC acting as a brake on the Federation had evaporated for good. Instead, the TLC actually assumed some of the characteristics of fellow-travelling, particularly when in August 1939 it endorsed an 'industrial' candidate for borough council elections. Recognised as an independently minded militant miner but not a communist, James Brown's municipal policy nevertheless included many of the issues on which communists had campaigned – a proper sanitation scheme for the town, a public library, council creche, a reduction in State Mine electricity charges (the State Mine power station also provided electricity for the town) and – ominously in McKenzie's estimation – a demand for the 'granting of equal political rights to all progressive political organisations… in attempts to increase the liberties of the people'. At the election, Brown headed the South Ward poll. Militant unionists had secured their base in the Miners Federation and now seemed ready to challenge the ALP over a range of political and social issues.

---

28 *Sentinel*, 11 August 1939.

Chapter 5

Contraction and contradiction, 1939–1945

The opening years of the Second World War further dislocated the economic and social life of Wonthaggi. Two issues underpinned a general sense of unease: the continuing economic decline of the State Mine and the lingering aftermath of the Log campaign.

The State Mine had proved an invaluable asset during the First World War and promised to serve the same function again. Throughout the war years, John McLeish, as general manager, worked assiduously to maximise production, yet circumstances once more conspired against him. Neither his efforts nor those of the mineworkers themselves could overcome the mine's problem of underdevelopment, an essential legacy of the Depression. In the six years prior to the outbreak of the Second World War, average annual production exceeded 300,000 tons, a figure that slipped during the war to 250,000 tons, even as national production soared. Two problems, at once inter-related yet seemingly insoluble, drove this crisis. Chronic underdevelopment determined both the course and capacity of production: McLeish's Annual Reports from these years list an unbroken catalogue of obstacles frustrating efforts to open new seams. Following a personal inspection of the mine before the outbreak of war, Federation general secretary Bill Orr had written to McLeish: 'In my view, Western Area contains a seam of coal providing more difficulties, from the point of view of production, than any other coal seam being operated in Australia'.[1] There is no doubt that, as an experienced manager, McLeish had anticipated these difficulties and had planned throughout the 1930s to conquer them, but to no avail. Development of the Kirrak Basin had been seen as the key to any solution, but Kirrak would never achieve McLeish's original estimate of 900 tons daily production. Bedevilled by extensive faulting and equipment failure, daily production instead stagnated at 100–150 tons. By 1945 the

---

1   W Orr to J McLeish, 2 December 1937. State Coal Mine Collection, Victorian Public Record Office.

Kirrak pits had been 'temporarily' closed, awaiting new turbine engines and additional miners. The mine's 1943 Annual Report explained the related problem: 'The acute shortage of experienced labour has continued to have an adverse effect both upon production and development... the results [of recruitment] have been disappointing'.[2] While enlistment for war service might be an obvious source of this shortage, the mine's workforce continued to decline even after a 1943 embargo on employee enlistment. The size of the mine's workforce fell dramatically during wartime, from an average of more than 1300 at the outbreak of war to 870 in 1945–46. Retirement at 60, supplemented by the mineworkers' pension – both benefits of the prewar Log campaign – only intensified labour shortages, and a governor's order-in-council proved necessary to suspend such retirements for the duration of the war.

War revitalised the Labor Party, and its embrace of Australia's war effort contrasted with the industrial truculence and political antagonism of the communists. 'Today', Bill McKenzie told local technical school students in 1940, 'we are at war with a nation posing as Social Democrats'. The true role of social democracy, he continued, could instead be seen in Labor's commitment to the war effort and to the mobilisation of all resources to support national defence. Labor most truly represented Australian nationalism, and to that end ALP members led much of the early war effort in Wonthaggi. Although much of the commitment evident in 1914 could be observed, the level of jingoism did not return. Even so, win-the-war rallies, lectures and sermons became a feature of everyday life.

Single-minded support for the war came at a cost for some. Since the mid 1920s a small but significant number of Italian immigrants had gravitated to South Gippsland and to the coal industry. Some could be found at the State Mine, although more worked in the small private mine at Kilcunda. During the Log campaign, the Federation had printed its rule book and many propaganda sheets in Italian in a successful attempt to recruit them to the union. The tables now turned when, in response to Italy's declaration of war in June 1940, mineworkers struck in two Wonthaggi pits, refusing to work with Italians. McLeish temporised, unwilling to lose efficient miners but loath to jeopardise production. Police solved his dilemma a week later when, in a series of dawn raids, eight Italians were arrested and others removed from essential war work. By late 1940 the outward trappings of war such as the

---

2 *Annual Report of the General Manager of the State Coal Mine, 1942–43*. 1943. Melbourne: Government Printer: 4.

arrival of metropolitan evacuees, air raid precautions and the inevitable civic send-offs for volunteers had once again become commonplace, but economic sacrifice had not. There would be no downturn in the mining industry as in 1914–15. The war effort quickly soaked up the last of the unemployed in a town enjoying full employment for the first time in more than a decade.

As McKenzie and the ALP prospered, the Communist Party (CPA) struggled for direction. After its initial support for Poland following the German invasion in September 1939, the CPA swung behind the Soviet Union position that characterised the war as 'imperialist', with the result that R G Menzies, now Australia's prime minister, declared the CPA illegal in June 1940.

In Wonthaggi, local communists maintained their political position by virtue of industrial strength and through front organisations such as the Political Rights Committee, an organisation ostensibly devoted to a fair go for all and the lifting of political bans on the CPA. Curiously, though, its leadership appeared interchangeable with that of the local CPA branch, with leading identities including Idris Williams, Alan Opie, Bob Hamilton and Jock Irving. Federation mass meetings continued to provide them with a platform, while night letterboxing and social contacts also continued as before. However, the police raids of June 1940 netted more than enemy aliens: many homes of known communists were visited, and police detained Tom Currie, the well-known communist union official, while out letterboxing. Significantly, authorities dropped his case without prosecution. One former miner, when asked in 1975 the reason for such leniency, replied briefly, 'If Tommie Currie had been arrested, the whole mine would have stopped work', a response that once again reflected the successful integration of Wonthaggi communists into the local workforce and wider community.[3]

Wonthaggi communists had organised on a community rather than industrial basis. Despite Wonthaggi's significance as a mining centre, no State Mine branch of the CPA ever existed. Instead, a Wonthaggi branch existed from 1933, later joined by Wonthaggi 1 and Wonthaggi 2 as well as a Bass Shire branch – covering the satellite suburbs of North Wonthaggi and Hicksborough – and, finally, an Inverloch branch to the east. Miners and their families provided a majority of party members, with their political allegiance strengthened further by well-developed ties of kinship and work experience. The CPA in Wonthaggi had not grown out of the unemployed movement or from disparate groups lately radicalised by the Depression.

---

3   Interview with Harry Bell, North Wonthaggi, 16 May 1975.

Instead, it had expanded from a social base of postwar immigrant mining families – principally Scottish, English and Welsh – that by the onset of the Depression had successfully been integrated into the local community. Party members could boast a range of civic involvement, positions and achievements to rival McKenzie's: by the years of Party illegality they were not a fringe group but a recognised centre of community life. The CPA's initial opposition to the war found an outlet in the Federation's determination to continue to prosecute outstanding industrial claims from the Log campaign. In doing so, it also reflected a sense of dissatisfaction among the wider membership of the union that war was being used as an excuse to defer a number of reforms that were seen as long overdue. The claim for a 40-hour working week for all mineworkers remained outstanding, as did claims for retirement and pension schemes, still before a Commission of Inquiry. To complicate matters in Victoria, wage demands ranging up to 27.5 per cent for contract rates were still to be adjudicated. While the Log campaign had brought undoubted gains, many mineworkers still feared the return of the Depression, convinced that war would only provide a temporary interlude of prosperity.

In February 1940 Wonthaggi miners struck for a uniform 40-hour week, remaining on strike for three months. Even when work resumed, sporadic industrial action plagued the State Mine for a further 18 months, for there were many miners, Jack Goldsmith being one, who saw in the demands of a war economy the opportunity to force further concessions. Communists built on this sentiment. In a radio address during April 1940 Williams went to considerable effort to explain the recent three-month strike as unfinished industrial business from the Log campaign rather than communist-inspired sabotage. Strike action ceased after a number of Federation demands were submitted for arbitration. The amended award announced the following August angered miners: not only did a uniform 40-hour week remain as elusive as ever, contract rates increased by a marginal 5 per cent. But Wonthaggi miners had a further grievance. In delivering his determination, Justice Beeby had also reversed the decision of Drake-Brockman to consider the mining industry on a national, rather than regional, basis. Notwithstanding vehement union arguments to the contrary, Beeby endorsed the proposition that continuous work now offered by the State Mine required a different wage fixation principle to that operating in NSW districts, where existing rates included a loading for intermittent work. Victorian rates must consequently be lower, he argued. How could it be fair, Williams responded, for surface workers at the State Mine to work longer hours each week than their NSW counterparts yet receive nearly A£3 less? Significantly, McKenzie joined Williams in venting anger at the tribunal's decision, supporting the strike for a uniform working week, condemning federal government inactivity and demanding intervention instead to achieve an equitable outcome. The political benefit of his continuing support became evident when, as part of the settlement to the three-month strike, miners agreed, by a disputed vote of 319 to 204, to accept a condition confirming arbitration, instead of strike action, to settle disputes at the State Mine. Such co-operation went further in early 1941 when the Federation agreed to replace the three-monthly cavil for working places in the mine with an annual arrangement. Such an agreement meant that two days lost every three months to transfering equipment within the mine could be used for production.[4]

Such small concessions proved to be a sign of things to come. Operation Barbarossa, Hitler's June 1941 assault on the Soviet Union, turned the world upside down. Local businessmen could find themselves in agreement with the general secretary of the Communist Party who urged that 'efforts should be made to settle strikes by negotiation so that production of munitions should not be held up'. When the recently elected Labor prime minister, John Curtin, wrote to the Federation in November 1941 that: 'Two things are paramount with the Labor government, namely, justice for workers and victory in the war. The first, even if we get it now, would only be temporary unless the second were accomplished', he found a ready and amenable audience.[5]

With the entry of the Soviet Union into the war, the Federation executed a swift about-face. Production for the engines of war became a higher priority than economic justice. Such a transformation inevitably led to industrial tensions within the union; militancy was not like a tap, able to be turned off and on at the whim of union officials. Rank-and-file mineworkers, dissatisfied with working conditions or management behaviour, could no longer assume automatic support from the union hierarchy. To the contrary, having spent four years supporting aggressive industrial action in order to restructure the coal industry to the advantage of its employees, miners now faced condemnation from their own leadership for advocating strike action in defence of working conditions. It proved a difficult situation for the Federation's National Executive to manage, and played out in different ways in different districts.

In Wonthaggi, the local Executive readily embraced this change of policy, suggesting: 'An immediate conference with the management and the union

---

4   *Common Cause*, 18 January 1941.
5   *Common Cause*, 29 November 1941.

to determine upon the measures for continuity and increased production of coal, which can be secured without detriment to our members'.[6] No such conference took place. Control of production and methods of mining had been the jealous preserve of management since the mine's establishment, and a traditionally schooled manager such as McLeish needed more than a suggestion such as this, coming out of the blue as it did, to change a lifetime's industrial practice. Despite this initial rebuff, Federation officials persisted: if they were no longer to resort to strike action, some other form of dialogue with management would be necessary. This change in direction could be seen in the language adopted by communists who as recently as a year previously had called the union out on strike indefinitely. In the words of leading activist Alan Opie, 'production could be increased… by adjusting existing anomalies'. The Federation suggested the creation of a series of 'production committees' with 'mineworkers granted direct representation at all collieries in the voice and control of administration and production'. McLeish remained apprehensive, the more so when the union added the rationalisation of coal production and limited introduction of mechanisation to the agenda for these committees. Eight years previously, he had suffered the ignominy of being forced to accept union-controlled pit committees. Even if these had now become an accepted part of the State Mine's industrial scene, McLeish knew that they still served to consolidate union authority in the pits. There matters stood until June 1942, when the Federation and coal owners reached an accommodation with the federal Labor government to establish a national system of industry regulation. Known as the Canberra Code, this agreement confirmed increased federal government control of industrial planning and production, as well as confirming existing customs and practices. The Code endorsed the introduction of production committees, but as consultative rather than deliberative bodies. Within two months of the Code's adoption, production committees were up and running at the State Mine. The committees quickly became forums for suggestions related to increased efficiency and rational production, an ironic circumstance in light of Wonthaggi's past industrial history.[7] Once he had agreed to such committees, McLeish effectively exercised a controlling influence, using them to promote policies such as disciplinary sanctions against absenteeism

---

6   Miners Federation Victorian District. 1941. *Declaration of War Policy*. Wonthaggi: Miners Federation Victorian District.
7   See ACSEF Minutes, 18 January 1942; 25 April 1942; *Common Cause*, 18 April 1942; *Sydney Morning Herald*, 1 June 1942.

that barely a year previously the union would have fiercely opposed. Now, the unions deferred to the general manager.

With production committees increasing efficiency at the mine, union acceptance of conciliation and collaboration with management guaranteed continuity of production. Acceptance of such a partnership had its difficult moments on the union side: here the past five years had to be put aside, which was no easy task. In order to maximise production, the Federation had to consciously break with the past and, at its own risk, rupture the very rhythms and patterns of militancy that had become the essence of the union itself. At the very least, this new way of working would neutralise a long-practised and successful industrial strategy, almost certainly to the mineworkers' own detriment. Despite this, the repudiation of the new award by a mass meeting of Wonthaggi members in January 1942 stopped short of strike action, and even when union delegates returning from the 1943 National Conference (especially called to defuse rising tensions within the union) reported that the prime minister (in private meetings) had 'threatened to put the Military powers into the pits' if disturbances continued, Wonthaggi miners responded more in sorrow than in anger. Perhaps Curtin's commitment to establishing a complete 'survey of mining conditions embracing health, dust, ventilation and other matters' muted any dissent.[8] At times, the contradiction between current policy and former practice became extreme, as when one party of miners was dismissed in February 1944 for failing to 'make the min'. The Branch Executive failed to convince management to adopt an alternative method of managing difficult workplaces, and the failure of management to concede the point resulted in terse arguments within the Branch itself, with union officials as well as managers the target of rank-and-file criticism. And yet, despite such tensions, for three years after 1942 the union's Wonthaggi leaders successfully maintained the no-strike policy. During that period, no days would be lost to strike action. The few resolutions seeking to commit the union to strike action all failed, defused by Executive commitments to approach the federal government to assist in resolving the issues in dispute. Only a union Executive willing to compromise, supported by a disciplined rank and file, could have ensured such a result.

Unlike the Federation's NSW Northern and Southern districts, Wonthaggi's miners entered into a wartime truce with the State Mine's management and both parties stuck to the agreement. Harassed Federation officials visiting from the faction-ridden Northern District or the strike-

---

8   ACSEF Minutes, 18 January 1942.

prone Southern District confessed themselves amazed at the tightly disciplined unionism of the Powlett Branch. The arrangement worked, but both parties recognised that when circumstances changed, so would wartime industrial arrangements. While readily acknowledging the strategy's success – 'the management and pit committees became a guiding influence over our members, unwarranted actions against members for trivial breaches was [sic] replaced by a degree of tolerance and understanding' – Williams prepared the union for a return to more traditional antagonisms.[9] McLeish agreed, giving a similar verdict on his retirement in 1945. While accepting the benefits of co-operation between management and mineworkers, he expressed a firm view that, at the conclusion of global hostilities, stoppages and strikes 'would be forced' and confrontation would resume.

As the district facing the most intractable physical problems yet still concentrating on the necessary task of coal production in support of the war effort, Victoria defined the Federation's wartime policies. District President Idris Williams reaped his reward in 1947 when Federation members elected him general president. Rank-and-file members sought an equally tangible payoff in the form of a Commonwealth program of postwar reconstruction. Curtin's promise to consider union claims relating to health and working conditions had led to the creation of the Davidson Royal Commission, whose recommendations would prove instrumental in reshaping coal mining in Australia. But miners also looked to a more general agenda for reconstruction to rebuild mining communities.

Despite a Labor Party call for 'an outline of general principles of postwar reconstruction' as early as 1942, planning only commenced in earnest after December 1943, following Ben Chifley's newspaper series identifying full employment, social security and a stable international economic order as Labor's themes for postwar development. Chifley's vision received the Federation's wholehearted endorsement. War had created conditions of full employment in the industry for the first time since 1927 (since 1930 in Wonthaggi) and no-one relished a return to a bitter regime of unemployment and social suffering. The recently elected general secretary of the Federation, Harold Wells, warned of the 'bitter internal strife with increasingly big strike struggles, increasing accidents, collapsing mines and the wastage of a great national asset'.[10] As an alternative, he outlined the three assumptions that provided the basis for the Federation's postwar activism:

---

9   Idris Williams, quoted in *Common Cause*, 23 January 1943.
10   Harold Wells, Miners Federation National Secretary, quoted in *Communist Review*, November 1944.

1. Postwar demand for coal would remain at a high level, and as a consequence coal-owners could afford to meet miners' demands.
2. The federal government had promised a new era for mineworkers and mining communities in recognition of their war effort.
3. The pressures of wartime fatigue and restrictions, together with the need to modernise mining infrastructure, made the job of reconstruction all the more pressing.

Based on his analysis, Wells outlined the Federation's strategic priorities: the need to nationalise the industry; a levy on coal production together with a system of price control to bring necessary order to 'anarchic' production; and recognition of the Federation as a national union acting as such, rather than as a clutch of autonomous districts. In response, union districts and lodges developed their own local reconstruction programs for the industry and its dependent communities.

Communists provided the drive behind many of these visions of postwar life. In September 1944 the CPA reached a peak national membership of 23,000, and among other issues turned its attention to local government, seen by the party as a stepping stone to parliamentary representation. While the *Wonthaggi and District Postwar Reconstruction Plan for Happiness, Peace and Stability* appeared under the auspices of the Wonthaggi Trades and Labour Council (TLC), the final document was, in fact, a refined version of an earlier plan, *Immediate Problems and Postwar Reconstruction for Wonthaggi and District*, produced by local CPA branches some months previously. Here, reconstruction planning had to address a sobering fact: the State Mine, the lifeblood of the town, had been shrinking in every respect since 1929. Although it had survived the Depression, mass retrenchments and narrowing coal seams, each of of these factors had taken its toll on the mine's long-term future. This cycle of decline had to be broken: 'while there are still millions of tons of coal in the District, there is an urgent necessity to explore for greater coal resources. The danger of Wonthaggi becoming a ghost town… must be avoided at all costs by the exploitation of our natural resources'.[11]

The Plan called for increased exploration, the immediate expansion of the mine's workforce by 300, the development of the mine's workshops

---

11 Wonthaggi Trades and Labour Council. n.d. 1944–45. *Wonthaggi and District Post-War Reconstruction Plan for Happiness, Peace and Stability*. Wonthaggi: Wonthaggi TLC; Australian Communist Party Wonthaggi District Committee. 1944. *Immediate Problems and Post-War Reconstruction for Wonthaggi and District*. Wonthaggi: ACP Wonthaggi District Committee.

as a central repair depot for all Victorian mines, and the creation of a state instrumentality to manage all Victorian coal mines, 'both shaft and open-cut'. Yet, significantly, the Plan also canvassed non-mining options. Wartime decentralisation had attracted textile manufacturers to Wonthaggi, and building on such initial steps, the Plan proposed railway construction workshops, light-engineering facilities and an aluminium plant, as well as canning and dehydration facilities for processing vegetables and seafood. In a more detailed, or at least more ambitious, form, the Plan harked back to an earlier vision of Wonthaggi as 'the Newcastle of the South', but with a crucial difference. Secondary industry would no longer exploit the town's mining base but instead serve as an alternative for future development.

The TLC had also turned its attention to Wonthaggi's social amenities, and necessarily so. The privations imposed by a war economy had only served to exacerbate the social and community cost of prolonged depression and the years of grudging recovery that followed. Wonthaggi had lost its position as one of the state's larger and expanding regional centres and was being outstripped in growth by other towns across the state and in Gippsland. No significant buildings had been erected in Graham Street since the Union Theatre 20 years previously.

In fact, the static nature of the town's infrastructure attracted the attention of the regional demographers A and J Macintyre, whose classic wartime social survey of country Victoria sought to help define the policies that would need to be addressed by any program of postwar reconstruction. The Macintyres' clinical dismissal of Wonthaggi's future reflects the ground the town had lost since 1930:

> Less money is invested in a mining town and houses and public utilities alike are poorer than they would [otherwise] be… For example, [at] Wonthaggi, where mines are already on the decline, there is a very noticeable lack of good building, either private or public.[12]

There were compensations, to be sure. The Macintyre survey noted Wonthaggi's unique public dental clinic and dental benefit scheme, as well as the beneficial social impact of the co-op and its influence in keeping prices for basic commodities low; but even such advantages, important as they were, could not disguise the fact that in the 40 years since 1909, the original ideal of a model state-run town had deteriorated into the depressing reality

---

12   A J and J J Macintyre. 1944. *Country Towns of Victoria: A Social Survey*. Melbourne: Melbourne University Press: 62.

of a weatherboard canvas-town, its permanence only serving to intensify the frustration and deprivation many mineworkers and their families felt. Housing in the town was, at best, adequate, but more often shoddy and in need of repair. Although mineworkers proved adept at improvisation and improvement, the accumulated impact of years of depression and short-time had denied them the resources necessary for adequate rebuilding. By 1945, only 115 of the town's 1200 dwellings met the Housing Commission's spartan minimum standards. Sewerage was often inadequate and some houses lacked a bathroom. The shortage of accommodation in the town meant that attempts to recruit additional labour drove rents even higher. As *The Sentinel* reported in February 1945, 200 dwellings were fit only to be demolished, yet an additional 500 homes would be required to house the population adequately.[13] The Reconstruction Plan envisaged state assistance in solving this housing problem, supplemented by provision of a library, swimming pool and increased parkland.

Achievement of such a grand vision depended on state intervention and to this end the Federation sought to enlist the support of the Borough Council, state parliament and sympathetic parliamentarians to the cause. Thirteen unions, the TLC, three political parties, the Wonthaggi District Hospital Committee and four citizens' associations – all with direct or indirect affiliation to the Federation itself – subscribed to the Reconstruction Plan. Unanimity in local support would be essential if the state government was to be convinced. Few governments, of any political hue, had been great admirers of Wonthaggi in past years. But regardless of any bickering with politicians, the Plan ultimately foundered on Wonthaggi's traditional curse – hostile geology.

Although everyone had been acutely aware of the mine's continuing deterioration, no-one anticipated the collapse in production after 1943. The State Mine had entered the war with four productive pits, but in October 1945, management, lacking the necessary machinery or the manpower, gave up the unequal struggle in the Kirrak Area, followed in December 1946 by the final closure of the Eastern Area, where only pillar extraction (the removal of the last coal following the completion of mining operations) had sustained operations through the war's final years. The Northern Area, too, increasingly relied on such pillar extraction, leaving the difficult Western Area as the mainstay of production. From about 200,000 tons in 1945–46, production fell dramatically to less than 130,000 tons four years later. The State Mine could no longer operate as a strategic reserve for the Railway

---

13   *Sentinel*, 23 February 1945; 23 May 1945.

Commissioners. At best, the mine supplemented coal imports from NSW. The material influences delivering this final permanent decline were not new. Faulting, stone bands, thin seams: hostile geology had proved an obstacle for 40 years. Finally, geology had won. Managers now lacked even the necessary manpower to throw at the problem. As the size of the workforce continued to fall, only continuing retirement prevented the average age of the workforce from exceeding its abnormally high figure of 42 years.

A little of the character of early 'tent town' returned, as a number of temporary huts and shanties had become permanent, if near derelict, fixtures. And in a reversal of the practice of 40 years previously, higher rents available in Korumburra provided sufficient incentive for the removal of local houses to the dairy town 20 miles to the north. Following months of constant agitation, the Housing Commission commenced construction of 25 houses in May 1946. By then, the situation had become critical. Two months previously, the last house advertised for rent had attracted 55 applications. Further Housing Commission projects commenced in 1947, with up to 100 families balloting for each house as it became available. As with so many other communities, Wonthaggi wanted and needed more housing, but the most direct advice local advocates received came from the Liberal–Country Party Minister for Decentralisation, Herbert Hyland: 'forget your politics and I think we can do something'. His offer was ignored.[14]

By these postwar years, active involvement in socialist activities had also brought Wonthaggi into the orbit of Australia's cultural renaissance. Wonthaggi's leading union officials now regularly rubbed shoulders with socialist intellectuals. Writing in September 1944 to congratulate Williams on his recent local government success, the noted historian and political commentator Brian Fitzpatrick sought Williams's assistance for his 'proposed short history of the Australian people', recalling that

> on my only visit to Wonthaggi I heard something about long underground treks, under low roofs that miners there had to make to their sections of the coal face and the bad conditions they found when they got there. Now, I want to include that sort of thing in my book.

Ironically, in light of the later 1949 strike disaster, Williams interleaved Fitzpatrick's letter in a copy of *The Mineworkers' Future*, a joint ACTU/Miners Federation pamphlet exhorting greater coal production with a

---

14 *Sentinel*, 9 April 1948.

Figure 41. Noel Counihan, *Miner*, linocut, 1947; one of six images from Counihan's Wonthaggi mineworkers series
Private collection

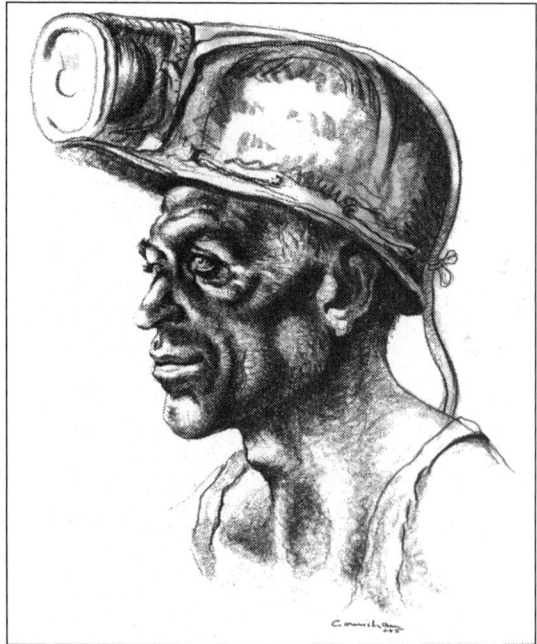

Figure 42. Noel Counihan, *Untitled ('Wonthaggi miner')*, pencil and conte chalk, 1942–43
Private collection

Contraction and contradiction, 1939–1945 | 129

Figure 43. Noel Counihan, *The young wheeler*, linocut, 1947
Private collection

Figure 44. Noel Counihan, *In the narrow seam*, linocut, 1947
Private collection

promise of a role for the Federation in the postwar restructuring of the coal industry and its dependent communities.[15]

Earlier that same year another friend of Fitzpatrick's, the communist artist Noel Counihan, had also visited Wonthaggi, similarly drawn by his need to understand the miners' fortitude under harsh conditions and his wish to translate their war effort into a series of paintings. Counihan lived with a mining family – Harry Webb and his wife – and worked with the miners, recognising that to successfully achieve his project he needed practical experience: 'I crawled wherever they crawled. I know that in the various seams, for example, the length of timber needed to prop up what they'd call the ground, we'd call the roof, were only eighteen inches long'.[16] From his Wonthaggi days Counihan produced a memorable series of works. In an industry familiar with graphic images, his Wonthaggi paintings, drawings and linocuts stand out. His drawings of individual miners capture both their resoluteness and their harsh years of experience, while his postwar linocuts are brutal examples of the industry's dangers and its victims. Above all, his paintings from the State Mine, particularly *Miners Working in Wet Conditions, Wonthaggi*; *Miners Preparing a Shot* and *In the Eighteen Inch Seam, Wonthaggi State Coal Mine* capture the inherent dangers of mining in confined spaces where elemental forces of compression and explosion created harsh, unforgiving conditions that distorted and reshaped the human condition. *Miners Working in Wet Conditions* deservedly won First Prize in the *Australia at War* exhibition that opened at the National Gallery of Victoria during the first heady months of peace in 1945.

On another front, Wonthaggi had cause for celebration in July 1948 when, 20 years after *The Sentinel*, together with councillors McKenzie and Easton, had begun to agitate to bring secondary industry to the town, Miller-Cyclone Forgings established a light-engineering metal fabrication plant in Wonthaggi. Housed in the State Mine's obsolete 18-shaft workshops, now relocated adjacent to the town's railway yards, Millers believed that it would employ 80 men within a year. Such an operation could hardly compare with the State Mine, which in its prime produced 600,000 tons of coal annually and employed a workforce approaching 2000. But those days had long since passed. Millers now pointed at an alternative path to the future.

---

15   Brian Fitzpatrick to Idris Williams, 17 September 1944. Private collection.
16   Quoted in Bernard Smith. 1993. *Noel Counihan: Artist and Revolutionary*. Melbourne: Oxford University Press: 195.

Chapter 6

'Back the Miners' Programme'

Isolation and confrontation, 1945–1949

Even if the State Mine had lost most of its economic relevance, the final major act in Wonthaggi's political history as a mining town had yet to be staged. With the State Mine's circumstances ebbing each year, what strength the Victorian Branch retained depended on the political influence of the Miners Federation nationally and the continuing success of its industrial strategy, a strategy to which Wonthaggi's miners had contributed so fundamentally. The tumultuous years between 1945 and 1949 would determine how far the Federation could carry its success following the return of peace.

During December 1945 Jack McVicars retired as Victorian secretary after an unbroken term of 32 years. In the Executive reshuffle that followed, the position of full-time president was abolished. Idris Williams, having served 12 years as president, successfully transferred to the postion of secretary (now the only full-time Victorian District officer), while Bob Hamilton became president. For the first time in Victoria, communists occupied both of the Federation's senior positions, a reflection of the Communist Party's grip on the union nationally.

In February 1946, at the first District committee meeting following his election, Hamilton outlined the essentials of the Federation's postwar strategy, and did so in uncompromising terms:

> We are about to start on a new era for this district, talking with the management for production is over. Now it is a question, as Wells [National Secretary following Bill Orr's retirement] stated, the winning of conditions comes first and in securing conditions we must play our

part in [the] labour movement... if the Labor Party does not assist us in improving conditions, then we will have to fight them...¹

Hamilton's calculated belligerence echoed the position taken by other, more prominent communists. A month previously, the state organiser for the CPA in Victoria, Sam Samson, had told an aggregate meeting of local party members that the national economy faced a period of dislocation as unions, now freed from the constraints of war, pressed urgently for overdue concessions, in particular an increase in the basic wage, the achievement of a general 40-hour working week and guaranteed, permanent employment. Implicitly, he argued that neither employers nor the Commonwealth Government would concede such claims willingly. A year later, Edgar Ross, editor of *Common Cause* and strategic advisor to the Federation's leaders, developed Samson's argument in a way that suggested a confrontation could not be postponed for too long when he told the Communist Party's Central Committee that 'to-day we are in a position of Governments, and particularly the federal government, being our most direct enemy and not the coal owners'.²

The Federation's 1943 National Convention had committed the union to a policy of nationalisation of the coal industry but, in the face of what the union identified as Commonwealth Government intransigence, it developed an extensive new Log of Claims that were ratified at the 1948 Convention. These included:

- a 35-hour week for all mineworkers: productivity was rising and a shorter working week was necessary to ensure full employment
- long-service leave: previously granted to Victorian State Mine employees, all Australian coalminers deserved this benefit
- a general 30 shilling increase in wages
- improved pit and town amenities to overcome two decades of underinvestment.³

The years 1945 to 1947 brought Wonthaggi's miners significant gains: an expanded Compensation Act – considered the only workable one in the industry, a satisfactory pension scheme, long-service leave (six months after 20 years), and successive wage increases. Such conditions placed Victoria,

---

1 ACSEF Minutes, 13 February 1946.
2 Edgar Ross quoted in Communist Party of Australia Central Committee Plenum Reports, 14–16 February 1947.
3 Miners Federation. 1948. *Back the Miners' Programme*. Sydney: Miners Federation.

however small in numbers, in the vanguard of Federation campaigning. In achieving this position the Federation had an indispensible ally, the state's Cain Labor government. In the union's own words, between 1945 and 1947 this government made 'peace in the mining industry' possible.[4] The amity between Wonthaggi miners and their state government could not be reproduced at a national level where, as predicted by the union's national leadership, the Davidson Royal Commission into the Coal Industry refused to consider nationalisation as a serious question, and the federal government itself rebuffed Federation proposals for legislation to introduce both a 40-hour working week and a weekly basic wage increase of one pound. While the active support of the Cain government had materially assisted the Federation's Victorian Branch, no amount of state government support could supplant, or even deflect, the intent of the Federation's national industrial campaigns. On the contrary, both John McLeish and his successor as general manager, Bob Johnston, complained that such government support actually confirmed the union's strategy rather than altered it. Even so, the effects of such collaboration were undeniable: lost time dropped away and working conditions improved.

In the overheated industrial relations climate of these postwar years, even such collaboration could prove to be a game of bluff and counter-bluff. Both the management's and the union's tactics would be framed accordingly, as was the case in a February 1946 dispute over a minor payments issue. ALP member and District Vice-president Jim Birt moved that 'the executive immediately wait on the Minister for Transport and failing complete satisfaction a stop-work meeting be held… and we remain idle until complete satisfaction is attained'. Also present at the meeting, by invitation, was the Minister for Agriculture and Mines Bill McKenzie. His response demonstrated the immediate strengths and longer term weaknesses of this strategy. Urging outright support for the resolution, McKenzie declared:

> As long as the Labor Government is there we intend pushing ahead with legislation that affects the interests of the workers. We may be kicked out at any time but as long as we are there we will get on [with] the job.[5]

But in committing the government to parallel action, McKenzie failed to refer even once to the sting in the tail of Birt's resolution, the threat

---

4   Quoted in the *Powlett Express*, 7 November 1947.
5   ACSEF Minutes, 6 February 1946.

Figure 45. Wonthaggi miners and family members march in the May Day demonstration, Melbourne, c. 1947
Members of the Miners Women's Auxiliary include Nancy Stirton (pushing daughter Anne), Meg Foster and Agnes Doig carrying the banner, and Elsie Hamilton on the left in the light-coloured suit.
Wonthaggi Historical Society Collection

Figure 46. International Children's Day celebrations organised by the Miners Women's Auxiliary, 1952
Wonthaggi Historical Society Collection

Figure 47. Members of the Women's Auxiliary at May Day, Melbourne, c.1947
Facing the camera are (from left to right) Elsie Hamilton, Nancy Stirton and Meg Foster.
Wonthaggi Historical Society Collection

Figure 48. Again at May Day, 1947
Union activist 'Andy' Williams stands on the left, with Auxiliary members Meg Foster and Nancy Stirton (and daughter Anne in the stroller).
Wonthaggi Historical Society Collection

Figure 49. Wonthaggi mineworkers gather outside the Trades Hall in Victoria Street, Melbourne, prior to a May Day demonstration in the late 1940s
CFMEU Mining and Energy Division National Office

of an indefinite stoppage. McKenzie's response assumed continuing co-operation and failed to acknowledge that with another, less sympathetic government, mineworkers could well be forced to take industrial action under circumstances far less favourable than those prevailing in 1946. In the same way, the Federation ignored the full implications of its strategy, particularly the assumption that despite any strategic or tactical divisions between them, a federal Labor government would ultimately support union objectives. This belief apparently held firm among the Federation's membership, despite increasing disenchantment on the part of a number of militant unions with the actions of the federal government. With the defeat of the Cain government in Victoria, this strategy became increasingly unstable and many would come to interpret political non-cooperation in terms of betrayal.

Although the town, due to its industrial militancy in the 1930s, had become known to some as 'Red' Wonthaggi, in organised politics it had always remained staunchly Labor. Over a period of more than 30 years a close relationship between the town, the State Mine and the ALP had come to be expressed through the person of Bill McKenzie. During that time he had consciously worked to be seen as the personification of industrial Wonthaggi, of its future in mining and of its unswerving allegiance to Labor. His appointment as a minister in Cain's administration represented the crowning success of his long career. But from the time of the Depression, or at least recovery, local ALP organisation had begun to contract. Communists occupied union positions formerly held by McKenzie's supporters, and a further consequence of McKenzie's importance to Wonthaggi had been the transformation of the local ALP into an organisation concerned with his electoral success.

By 1945 Labor's grip on the state seat of Wonthaggi had been weakening for nearly a decade. Following the decision in 1947 by a Liberal-controlled Legislative Council to deny the Cain government supply, McKenzie went to the people for an eighth time. As a regional constituency, Wonthaggi had never been large in terms of voters, a convenient fact that allowed 3000 electors in Wonthaggi and, to a lesser extent, a further 1500 or so in Korumburra, to dominate the electorate. The decline of coal mining had influenced events far beyond the boundary of the State Mine. As the electoral balance shifted away from Wonthaggi, so did the allegiance of at least one local paper, the *Powlett Express*, which had abandoned its traditional Wonthaggi focus to act as the voice of the town's rural hinterland. It proved to be a disenchanted voice:

> Promises, promises, promises… Wonthaggi, a Labor town, rejoiced when the present Labor government came to power. Voters thought that at last they would get a go. But they haven't.[6]

It seemed possible that McKenzie's deliberate identification with Wonthaggi and with mining could now prove to be a handicap. The *Powlett Express*'s 'plague on all houses' rhetoric did, indeed, reflect a growing anti-Wonthaggi sentiment. Some complained that while Wonthaggi received new housing, surrounding towns only received Wonthaggi's cast-offs. Others expressed a commonly held view that 'communist' miners received preferential treatment from the Cain government. At the November election, McKenzie's Country Party opponent swept all but six mining booths – Wonthaggi, Wonthaggi North, Dalyston, Dudley, South Dudley and Kilcunda – outpolling McKenzie by more than 2000 votes.[7]

Such a comprehensive defeat inevitably led to recrimination and attempted retribution. Militant members of the ALP, such as Jim Birt (now District

---

6 *Powlett Express*, 21 February 1947.
7 Figures complied from *Sentinel*, 14 November 1947; 28 November 1947; *Powlett Express*, 14 November 1947.

secretary following William's election as national secretary), shared control of the Victorian District with communists. Their obvious preference for such collaboration found little support from McKenzie or many other party members. In the aftermath of the 1947 electoral defeat different political strategies flared into open dispute, with Birt and Jim Stewart, an ex-committee member and active unionist, charged with being secret Communist Party members. A branch inquiry rejected the charges, but the fact remained that Birt and Stewart, like a number of other influential Federation members, did interpret their ALP membership in ways very different to McKenzie.

The 1949 general strike has too often been considered as either a communist conspiracy or a strike by mineworkers goaded beyond the point of endurance. Neither interpretation is adequate. It is true that there would have been no such strike without communist direction of the union but, equally, mass meetings of mineworkers endorsed the strike strategy and were clearly of a mind to accept confrontation, as Wonthaggi's miners had been in 1934. The strike represented a conscious, planned escalation of Federation strategy (much along the lines of the 1937–40 Log campaign) yet no analysis can ignore how ill-conceived the strategy proved to be, and the relative ease with which the strike would be isolated.

After 1947 the Victorian Branch had supported a number of successful militant-led strikes in Victoria and interstate. The industrial gains achieved by railway workers in Victoria had been eclipsed later in 1947 by the success of Queensland railway workers after a nine-week strike, and both strikes appeared to confirm the effectiveness of militant industrial action. At the same time, the federal Labor government, in power since 1942, was beginning to show signs of wear and tear, and debts owed to the union movement from the days of war had well and truly been paid. The federal Labor government no longer needed to placate the Miners Federation, and if the union was to challenge its authority, the response would not necessarily be what the Federation might expect. In other words, any challenge to the Chifley government would be resisted with all the force at the government's command.

Unlike in 1934, Wonthaggi would never be at the centre of the strike in 1949. Above all, this would be a confrontation decided on the Northern coalfields of NSW. Yet Wonthaggi miners could never be dispassionate observers. In the months leading up to the strike, a campaign to equate unionism with communism came to dominate political affairs. The conservative Australian Constitutional League had begun to advertise in local papers, suggesting that Wonthaggi miners were complicit in a plot to 'destroy us as a nation'. Its claim reflected a process of political polarisation that threatened not only the Federation's own assumptions but the existence of the federal Labor government. Throughout March 1949, the local press also carried advertisements sponsored by the Friends of the Soviet Union (FOSU) announcing a lecture night at the Union Theatre, at which John Rodgers, FOSU secretary, was to speak of his recent visit to the Soviet Union. Regarded by many as a provocateur, Rodgers had earlier met with organised resistance at Shepparton, and the same thing now happened in Wonthaggi:

> A terrific demonstration smashed the John Rodgers meeting on Russia… Nine policemen were stationed in the theatre in case of emergency – they did not interfere with the demonstrators… the demonstration was organised on a wide scale, with servicemen coming from as far afield as Korumburra, Leongatha and Cowes.[8]

The only parallel to this meeting in Wonthaggi's history had been the violent anti-conscription rallies of 1916–17.

It proved a singularly successful demonstration. All attempts to establish order proved useless; preliminary speakers were hooted at and shouted into silence, and pandemonium prevented Rodgers from speaking at all. The following day, State Mine employees stopped work to hear the talk Rodgers had been unable to deliver the night before, and as a consequence of the demonstration the Federation ceased advertising in the *Powlett Express*. The Women's Auxiliary carried its opposition a step further, picketing the office of the *Powlett Express* while boycotting the businesses of those identified with the RSL, which had been associated with the demonstration. But demonstration and counter-demonstration aside, the fact remained that Wonthaggi's hinterland had been successfully politicised and had symbolically conquered the bastion of mining unionism – the Union Theatre – on 30 March.

Following this debacle, R G Menzies, already well launched on his drive to regain the Lodge, spoke in Wonthaggi during his well-publicised 'confront the miners' tour of April and May. Hecklers interrupted his Wonthaggi speech, but he spoke nevertheless, informing an unconvinced audience that in a 'classless society' unions had no justification in tolerating communist leaders. Seemingly in response, the Wonthaggi RSL expelled all communist members, as if incapable of true allegiance. By the time of the 1949 strike, politics had already sorely divided Wonthaggi.

---

8   *Sentinel*, 1 April 1949.

Organised along the lines of the prewar Log campaign, the Federation's industrial campaign climaxed years of postwar crisis in the industry. Federation claims were initially rejected then referred to arbitration. In years past this would have sufficed to temper any union response, but the Federation's leaders were no longer in a mood to conciliate with Chifley's Labor government. This strike represented the ultimate test of 15 years of communist leadership, and despite the polarisation of society and the growing divide between mining unions and the federal government, the Federation's communist leadership and its rank and file remained confident of success.

The tabling of the Federation's final Log of Claims in February was overshadowed by the Victorian Coal Tribunal's response to earlier union and management demands. The union interpreted the management's proposals as a prelude to wage cuts of up to one pound per week. While rejecting the management's case, the tribunal chairman, Mr Ramsay, agreed that earlier increases had proved too generous. But by proposing to cut the rates for general classifications by up to ten shillings each week and reduce shift rates by six shillings and six pence per week, he appeared to confirm all the Federation's worst fears. Faced with the threat of a general stoppage after 5 May, the State Mine's management agreed to a resumption of work under the previous award, while agreeing to consider the Federation's national Log of Claims. Although satisfied that another assault on conditions had been repelled, Wonthaggi mineworkers continued to insist on the 16 June deadline set for the acceptance of the Federation's demands.

In the shadow of the Federation's ultimatum, Williams told an overwhelmingly sympathetic stop-work meeting at Wonthaggi on 16 June that 'we are not asking for a general strike. If miners demonstrate that they are united and strong, I think that the authorities will settle [our claims] within a week'.[9] Few others were as optimistic. The next day, and only by a bare majority of 8–7, the Federation's National Council rejected a resolution put to it condemning the communist strategy, particularly its opposition to arbitration. Despite dissent at the top, communist determination to stage a decisive strike had, if anything, strengthened, and in this they enjoyed the overwhelming support of the union's membership. At a series of aggregate meetings on 23 June, the union voted to strike indefinitely by a margin of 7995 to 822. In Wonthaggi, only a single dissident prevented a unanimous vote.

---

9   Quoted in *The Age*, 17 June 1949.

The general strike isolated Wonthaggi from the Federation and from the rest of Victoria. In this respect it was no re-run of 1934. By mid July the town's rail link with Melbourne had been reduced to a single daily goods train. Few owned cars and, in any case, petrol remained rationed. Only a limited number of union speakers and organisers left Wonthaggi, with the majority of the population confined to the town. No other strike sparks in the memory of miners and their families an atmosphere of claustrophobia as strong as this one, a consequence perhaps of the town's isolation as well as of the State Mine's continuing decline. On 25 July, the State Mine closed its power plant. As long as miners refused to mine coal for the power plant, the state government would not connect Wonthaggi to the Electricity Commission grid. Overnight, the conditions of 30 years earlier returned: wood stoves, chip heaters and kerosene lamps.

Of course, the Federation had its own resources. Preparations had not been neglected and Wonthaggi's well-established Broad Committee mobilised to distribute food and other essentials. The 'gunny-bag' parade returned. Contributions from Melbourne sympathisers supplemented local sources of fresh food:

Figure 50. Three Wonthaggi miner leaders during the 1949 strike
From left to right sit Bob Hamilton, Idris Williams and Jim Birt, with an unidentified Tasmanian member on the right.
Wonthaggi Historical Society Collection

Figure 51. Members of the Wonthaggi mineworkers' Propaganda Committee on the steps of Unity Hall, Bourke Street, Melbourne, during the 1949 coal strike
Agnes Doig stands on the left, her husband Wattie Doig stands third from the left, while Bob Russell and Bill Stirton stand on the right of the picture.
Wonthaggi Historical Society Collection

> Joe Foster and Tom Bennet, together with our band
> And all Young Miners of the League, they gave a helping hand
> Emrys White, Jim Hudson, Fred Brown and Joseph Clough
> All helped unload the Melbourne trucks to distribute the stuff.[10]

Each week, families again queued down the lane running beside the Union Theatre to receive rations of meat and vegetables, butter and milk, tinned food and clothing. The arrest of union leaders and the court-ordered freezing of union funds in July only served to reinforce such principles of self-sufficiency. Despite the stationing of a security officer from the Federal Police in the town, Wonthaggi became a source of much-needed finance for the union nationally. Local donations supplemented the union's own funds, much of which was now held as cash, hidden by trusted union

---

10   Untitled, undated verse (1949). Miners Federation Collection, University of Melbourne Archives.

members and supporters behind pictures, in the stuffing of sofas, wrapped in oiled paper and pushed down drainpipes or cached under floorboards. Some of the money was used locally for strike relief, handed out in the Union Theatre with elaborate precautions taken to prevent its seizure by the local police. Other funds went surreptitiously to Melbourne via union and Communist Party couriers. When added to metropolitan collections, this money would then be smuggled to NSW for distribution throughout hard-hit mining districts.

The intensity of the government's reaction stunned many. While Victorian minister Wilfred Kent-Hughes suggested that the closure of the State Mine might be an inevitable consequence of the strike, vigorous community reaction ensured a swift retraction. But it would be the reaction of the federal government – and a Labor government at that – that sparked most outrage: 'never in the history of the labour movement', complained Birt, 'has anything worse been dealt out. We could not have expected worse from anyone'.[11] Many were willing to give him an answer. They questioned his commitment to the labour movement, along with the wisdom of rejecting arbitration and striking at such a critical time so close to a federal election.

The Wonthaggi miners' outrage scaled new heights following the arrest of Williams, together with other leading officials. They agreed not to return to work until his release. Such jailings and the sequestration of union funds unbalanced the campaign, while the introduction of the *National Emergency (Coal Strike) Act* enabled the federal government to use troops to work open-cut mines in NSW. In an attempt to regain momentum, Birt approached fellow members of the Wonthaggi Borough Council on 18 July, seeking their endorsement of a resolution protesting the federal government's introduction of its emergency legislation. Birt's approach left no councillor room to hide: 'this Bill is a war on women and children to crush the unions into submission... If I am to be torn between the Labor Party and my union principles, I will stand behind my union principles'.[12] His advocacy succeeded. Council agreed three to one, with a single abstention, to forward such a protest. Only John McKenzie, Bill's son, dissented. He argued instead that all the matters at dispute could be resolved through arbitration. This opinion carried considerable weight, as John McKenzie was also secretary of the local branch of the Engineers

---

11 *Sentinel*, 22 July 1949.
12 *Sentinel*, 22 July 1949.

Union. In the wake of Birt's action, the Amalgamated Engineers Union (AEU) backed its secretary, condemning the Federation's actions.

At a mass meeting on 1 August called to determine rank-and-file support after five weeks of strike action, more than 400 members rejected any return to work other than on Federation terms. But during the following week, union resistance collapsed on the Northern NSW fields. On 10 August, in a national ballot, Federation members now voted to return to work pending negotiations with employers on union claims. Victoria stood out as the only district voting to continue the strike. Williams had not yet been released. The intransigence of local Federation members only served to emphasise the degree to which local politics had changed. The Federation's total defeat in 1949 completed the political disintegration that economic decline had begun, a decline confirmed by McKenzie's electoral defeat in 1947.

Chapter 7

'When the last wheel finally turns...'
1949–1968

Total defeat in 1949 numbed Wonthaggi mineworkers. Although Kent-Hughes's threats of mine closure had quickly been dismissed, it would soon be evident that neither the State Mine nor its workforce would escape the consequences of 1949 unscathed. The defeat confirmed the demise of coal mining in Wonthaggi, further isolating mineworkers within an increasingly rural and agricultural community.

For 15 years, the Federation's industrial strategy of organised, focussed industrial militancy had met with almost unbroken success. It failed in 1949, and Idris Williams's justification, that 'the strike was a victory for organisation, and if it had not been for the Fifth column among the leaders of the Northern fields we would not be getting it in the neck today', had the same tone of self-justification that had been used in 1930 by the very Federation leaders that Williams had worked so hard to replace.

Crisis followed, within both the Federation and the ALP, while the Communist Party continued to lose ground in its trade union strongholds. In Wonthaggi, the ALP finally moved to prosecute its own 'intractable' members. For 40 years industrial militancy, and those who practised it, had held a secure, if contentious, place within the ALP. Their strength stemmed from the influence that their unions exerted within the party, and from the acknowledged importance of trade unions as the industrial wing of a broader labour movement. With the ALP in Opposition, or when a state Labor government such as John Cain's actively supported miners' claims, this partnership flourished. But when, as in 1949, besieged Labor governments battened down the hatches and sought to dampen industrial disputes in order to concentrate on pressing political issues, such alliances came under intense strain. The choice many such militant union members of the ALP had to make would not be an easy one.

Local borough elections in August brought political issues simmering since the strike to the boil. Already unpopular for leading the AEU out of the strike, John McKenzie was defeated by an Independent Labor candidate standing with Federation support. The Federation also actively opposed his simultaneous attempt to win ALP pre-selection for the state seat of Wonthaggi. 'I might add that a strike breaker is not likely to get the support of the Miners' Union', Branch President Bob Hamilton commented acerbically.[1] As retribution, charges of disloyalty to the party had been laid against a number of Federation members, most notably Jim Birt, the union's Victorian secretary, who now stood accused of being 'in the communist camp'. In many ways the charge stood – he was accused of being too close to the men with whom he had shared control of the Victorian Branch for many years. Accusations based on hearsay – 'Nine years ago I walked into a bord (a working place at the coal face) where Mr Birt and his son were working. They were talking politics and I was surprised to hear what I did as he was a leader of the ALP. He said he was a communist and the place for a communist was in the Labor Party' – could be refuted but parallel charges of disloyalty to the ALP, of supporting anti-Labor strike action and of supporting opponents of ALP-endorsed candidates, proved harder to dismiss.[2] Birt refused to recant, and the press report of the ALP meeting at which the charges were considered produced a classic exchange defining two opposing Labor philosophies in collision:

> Birt: 'The unions are the basis of the ALP and if the loyalty of my union is an offence then I am guilty… [McKenzie's] actions shocked me as he was a secretary of a union involved in a strike.'
>
> Unnamed respondent: 'This last strike was communist inspired… It was a strike against the Chifley government… The Labor Party is bigger than all the unions [and] there are too many members of the Labor Party who are fellow-travellers.'[3]

The meeting found the charges against Birt sustained and referred them to the state Executive for further action. This Birt forestalled by resigning, attacking again those 'who believe that by being loyal to a union, a crime is committed towards the ALP'. Likeminded unionists followed Birt out of the party. In November, James Brown, another independent-minded militant

---

1   *Sentinel*, 9 September 1949.
2   ibid.
3   ibid.

and a former borough councillor, resigned with a similar parting shot: 'if you do not think the same way as Mr McKenzie you are a communist'.

In November mineworkers re-elected the Federation's Victorian Executive unopposed. Local AEU officials, still deeply resentful of Federation attacks on their integrity, were also all re-elected. The previous month the Victorian District had disaffiliated from the ALP at the instigation of Birt and Hamilton. But such feuding had its limits. While mass meetings endorsed disaffiliation, miners proved reluctant to recommence industrial hostilities, rejecting in October Executive-sponsored calls for a sympathy strike in support of the jailed communist leader Lance Sharkey. The Federation found itself in an unfamiliar state of disequilibrium, with political disillusionment compounding the effects of industrial defeat.

The Labor Party's defeat at the subsequent November federal election set a pattern for ensuing years, years of struggle seldom relieved by success. The Federation itself had lost much of its industrial authority and hence its ability to effectively engage in the inevitable restructuring of the coal industry. The NSW Northern District proved to be the greatest loser. There, old mines would be abandoned, never to re-open. In coming years, both the Southern District and a rapidly growing Queensland industry would challenge its historical supremacy. Wonthaggi, too, was a loser. Amid the post-1949 themes of industry rationalisation and centralisation the State Mine could neither compete nor, ultimately, survive. The Victorian Branch of the Federation would be similarly affected. Wonthaggi miners would never again go out on protracted strike action. If local militancy served to preserve local conditions, this preservation followed negotiation rather than strike action. Coal-face disputes – an inevitable part of mining – continued to erupt sporadically, but the Branch's involvement was now simply industrial, bereft of ideological obligation. Like the State Mine itself, local mineworkers were industrially exhausted. The most unchanging aspect of local Federation activity would be its political stridency, its continuing championing of progressive causes and issues such as Menzies's ultimately unsuccessful attempt in 1951 to once more ban the Communist Party. But if general political settings remained relatively constant, the state of the industry did not. With overcapacity in the industry again jeopardising less productive mines, by the middle of the decade mineworkers were again preoccupied with the immediate prospects for the State Mine, but for a different reason than previously. After 1949 miners no longer focussed on prospects for expansion and diversification but, instead, the preservation of a diminished industry. As Alan Farmer, District president in 1953 explained:

We want if possible to finish our time in the pit, and enjoy Long Service Leave and pensions on reaching 60... Well, mates, to achieve this it is necessary for us all to pull together, give the Mine a product they can sell, cut down on absenteeism and achieve an economy that can be reached to allow the opening of Kirrak and Eastern Areas.[4]

With production once again in freefall and a permanent ban on further recruitment to the mine's workforce, members' concerns at continuing decline came to a head in May 1956 at a mass meeting that unanimously endorsed the need for a public meeting 'to discuss the best ways and means of retaining the coal-industry in Wonthaggi'. Significantly, traditional arguments relating to the strategic value of the mine had now been jettisoned, replaced by arguments concerning the value of the mine to its local community and the wider Gippsland region. Only when the argument turned to economic benefit did it falter: it was one thing to claim that the State Mine still represented a strategic asset, but another thing to demonstrate it.

The tactics adopted by the Federation demonstrated the limits of its case. The union no longer considered an industrial campaign to drive home its demands but instead sought consultation with the state government, ultimately co-opting the government's support for its case. While it went to considerable pains to deny any intention of running down the mine, the government's obvious caution and carefully chosen words exacerbated the very fears that it sought to allay. To ensure no back-sliding on the part of the government, a Federation-inspired public meeting in July 1956 established the Wonthaggi and District Vigilance Committee, charged with recruiting business and community support to ensure the continuing development of the Kirrak Basin; the resumption of further exploratory drilling work; and the attraction of coal-dependent secondary industry to the town. The Vigilance Committee sat firmly in the tradition of the public coal sale agitation of 1913–14 and the Wonthaggi Advancement and Defence League established by the Federation in 1935, but with a difference. This Vigilance Committee campaigned for survival rather than expansion. It succeeded in convincing the government to re-open the Kirrak operations – henceforth the mainstay of mine production – but could not convince ministers to agree to further exploratory work or to the recruitment of additional miners. The mine no longer possessed the capacity for regeneration. Henceforth, it would work to a negotiated close, and the debate on the future of Wonthaggi moved on.

---

4   Miners Federation, Victorian District. 1953. *Miners Union Bulletin*.

The Vigilance Committee campaign between 1956 and 1958 proved to be the Federation's last major campaign in support of an indigenous coal industry. Wonthaggi's mines had survived at first due to their strategic importance to the Victorian railway system and later by the industrial and political influence of the Miners Federation. By 1958, neither justification was compelling. The long-accepted nexus between coal production and Wonthaggi's prosperity had finally been broken, and even the souvenir pamphlet from the town's jubilee celebrations in 1960 confirmed this:

> Shortly after World War 2 came a body blow. The increasing use of fuel oil sounded the death-knell of the mines. The town had not foreseen this and was stunned... But there was still fight left in it. Secondary industries were attracted, others were begun by residents. These industries, coupled with the rich agricultural capacity of the district began driving Wonthaggi [again].[5]

This campaign also finally rectified what appears as a curious anomaly in the history of Wonthaggi as a stronghold of mining unionism. Drawing on strong British traditions, Australian coalminers have since the nineteenth century been at the forefront of commissioning union banners for both celebratory and industrial purposes. Yet it seems that for the majority of its existence the Wonthaggi Branch of the Miners Federation had no such banner to rally behind. It is true that its predecessor, the Victorian Coal Miners Association, commissioned a banner, and the terms and the circumstances of its commissioning are remembered by Jack McVicars. In remembering the aftermath of the bitter 1902–03 strike, he wrote:

> Many unionists left the [South Gippsland coal] field, through sheer starvation to seek work elsewhere... A number remained, determined to see it out, and the hardships they experienced could not be explained on paper.
>
> Our District Secretary at the time was Arthur Wilson. I remember, in addition to his sterling qualities as secretary, him organising concerts throughout the District in which he succeeded in raising funds to procure a Miners' banner which the miners marched behind at Melbourne during the '8 Hours' procession celebrations.[6]

---

5   Wonthaggi Jubilee Celebration Committee. 1960. *Wonthaggi Jubilee Souvenir.* Wonthaggi: Wonthaggi Jubilee Celebration Committee.
6   Untitled manuscript, autobiographical memoir of John (Jack) McVicars, n.d. Wonthaggi Historical Society.

This banner survives only in a few blurred newspaper photographs, and there is no evidence of it travelling to Wonthaggi with union members after the establishment of the State Mine in 1909. Similarly, there is no evidence of a newer banner being commissioned by Wonthaggi miners. Melbourne remained the geographical point of reference for Wonthaggi mineworkers celebrating the Eight Hours Day and no vibrant annual celebration ever developed in Wonthaggi.

The VCMA had commissioned its banner during an industrial and political crisis: it served as a demonstration of union claims to stability and permanence. At the other end of the District's history, the Vigilance Committee campaign had a similar result. After decades without a banner, the Wonthaggi Branch commissioned one, carrying it for the first time in the 1958 Melbourne May Day march. Designed and painted by a local sign-writer, Al Hannaford, the banner is bright red and depicts the stylised profile of a coalminer silhouetted against a headframe. In addition to the name of the union, the Wonthaggi Branch of the Australian Coal and Shale Employees Federation, the banner carries the uncompromising slogan: 'United for Socialism'. Like its VCMA predecessor, this banner was born of political adversity and the placard that preceded it on that 1958 Melbourne May Day march summed up the miners' position on the threatened future of their mine: 'Let Bolte Go First'.

But neither a new banner nor the miners' own determination could stem the ebb tide. Retrenchment of a further 100 miners in 1958 confirmed that the government did not intend to change its policy of a gradual yet inexorable wind-down to mining operations. Henceforth, the number of workers at the mine fell in line with production. By 1968, a rump of 140 miners produced just 33,000 tons of coal. This year would be the State Mine's last year, one preoccupied with arrangements for mineworkers' pensions or for alternative employment. The heat had now entirely gone from industrial relations. After 1960, only three days were lost to strikes, and only then in solidarity with Melbourne Trades Hall Council calls for industrial action. The elevation of the last general manager, Jim Byrne, from the ranks of the staff – he had begun work at the mine as a 15-year old clipper, ultimately becoming general manager in 1956 – exemplified the integration of staff and workforce during these years.

On 18 December 1968, Merv Campbell, the Powlett Branch's last secretary, wrote the Wonthaggi mineworker's farewell to mining:

> With the last pay about to be drawn and our last stump paid… we would like to convey to the office staff our appreciation for the help

and assistance which has always been extended to us in our efforts on behalf of our industry and the people who have depended on it for their livelihood. When the last wheel finally turns and comes to a stop, the days of reminiscing about the times when Wonthaggi was a big-gun mining town will begin.[7]

Byrne's reply, and his sense of history, proved equally acute:

> It is the closing of an era in Wonthaggi's history… Mining has been in our blood for many generations, and I have fond memories of the men who, through their work, not only laid the foundations of Wonthaggi, but of the development of this part of Gippsland.[8]

---

7   M Campbell to J Robertson, 18 December 1968.
8   J Byrne to M Campbell, 23 December 1968.

Bibliography

Notes

This volume draws heavily on my unpublished 1977 thesis, 'Industrial men: Miners and politics in Wonthaggi, 1909 to 1968', submitted for the degree of Master of Arts at La Trobe University, Melbourne. A more comprehensive bibliography, especially relating to the numerous political pamphlets consulted, is provided in that thesis, which can be consulted online or at La Trobe University.

The Archives of the State Coal Mine are held by the Victorian Public Record Office (VPRO) and the collection is a substantial one. At the time of my original research, the collection had yet to be fully organised and catalogued, and so material cited from the State Coal Mine Archive lacks file or reference numbers. Much of this work has since been completed and further details of the collection can be obtained from the VPRO.

Unpublished sources

Miners Federation
Cavil books: Bords, contributions and levies. 1912 to 1952. University of Melbourne Archives.
Correspondence between mine management and union authorities. 1914 to 1960. Noel Butlin Archives, Australian National University (ANU).
'Federal Convention: Second round of the Log'. 1937 to 1941. volume of cuttings, speeches, pamphlets, correspondence. University of Melbourne Archives.
'Minutes of general, special and stopwork meetings of members and committee of the Powlett River Branch, Wonthaggi' and later the 'District Committee of Management'. November 1911 to July 1954. Volumes 1 to 7, 11 to 15, 20, 21, 25 and 26 are held at the Noel Butlin Archives, ANU; volumes 8 to 10, 16 to 19, 22 to 24 and 28 are held at the University of Melbourne Archives. Volume 27, covering the 1949 strike, is missing.
'Minutes of meetings of combined unions of Carpenters, Engineers, Enginedrivers and Miners'. November 1917 to January 1935. Noel Butlin Archives, ANU.
'Minutes of meetings of the District Committee of management of the Victorian Coal Miners Association, later the Australian Coal Miners Association and Victorian and Tasmanian districts, Australian Coal and Shale Employees Federation'. May 1915 to February 1930. Noel Butlin Archives, ANU.
'Minutes of meetings of the State Mine Lodge'. September 1925 to May 1931. Noel Butlin Archives, ANU.
'Minutes of the Wonthaggi Branch Hall Committee'. July 1938 to May 1943. University of Melbourne Archives.

Theses

Cochrane, Peter. 1973. 'The Wonthaggi coal strike, 1934'. B.A. Honours thesis, Melbourne: La Trobe University.

Deery, Phillip. 1976. 'The 1949 coal strike'. Ph.D. thesis, Melbourne: La Trobe University.

Reeves, Andrew. 1973. 'The rise and decline of industrial unionism: The Workers International Industrial Union in Australia'. B.A. Honours thesis, Melbourne: University of Melbourne.

Reeves, Andrew. 1977. 'Industrial men: miners and politics in Wonthaggi, 1909–1968'. M.A. thesis, Melbourne: La Trobe University.

Miscellaneous

'Bill' McKenzie scrapbook. Newspaper clippings and photographs relating to McKenzie's parliamentary career. In the possession of the McKenzie family. Wonthaggi.

Knight, J L. 1956. 'Problems affecting the Black Coal Industry in Victoria'. Unpublished paper deposited with the Department of Mines Library, Melbourne.

McVicars, J (Jack). Unpublished reminiscences of a Wonthaggi miner and union official. Wonthaggi and District Historical Society.

Rankine Collection. A unique collection of political, industrial and social material relating to Wonthaggi and the State Mine. The collection includes rank-and-file miners' bulletins, union publications, photographs, job sheets and political paraphernalia, together with material published by the Friendly Society and Dispensary, the Workmen's Club, Union Theatre and Wonthaggi Hospital. University of Melbourne Archives.

Scott Papers. c. 1916 to 1923. Copies of correspondence between J B Scott and others involved in militant union and political organisations in Victoria, Western Australia and New South Wales. University of Melbourne Archives.

State Coal Mine, Wonthaggi Archives. 1909 to 1968. Inward and outward letter-books; fortnightly, monthly and annual mine managers' reports; personnel record cards; financial reports; private correspondence series with the Railway Commissioners; developmental and production files. Victorian Public Record Office.

Stirton, W (Bill). 'Depression in Victoria'. Unpublished reminiscences of a Wonthaggi miner. copy in possession of the author.

Wilson, A (Alf). 'All for the cause'. Unpublished memoirs of a syndicalist political organiser. University of Melbourne Archives.

Wonthaggi and District Historical Society. 1909 to 1968. Subject files on a range of issues and personalities connected to the activities of the Miners Federation, to its associated organisations and to union-related organisations in Wonthaggi as well as to the history of the State Coal Mine.

Wonthaggi Vigilance Committee. Minutes of meetings, 1956 to 1968; correspondence, 1956 to 1959. University of Melbourne Archives.

Workers International Industrial Union, Local No. 3, Wonthaggi. Minutes of meetings, May 1918 to February 1920. Noel Butlin Archives, ANU.

Published sources

Articles

Blake, J D. 1972. 'The Australian Communist Party and the Comintern in the early 1930s'. *Labour History* 23 (November): 38–47.
Cochrane, P. 1974. 'The Wonthaggi Coal Strike, 1934'. *Labour History* 27 (November): 12–30.
Dixon, R. 1970. 'Industrial policy in the 30s'. *Australian Left Review* 27 (October–November).
Edwards, A.; Baker, G; and Knight, J. 1944. 'The Geology of the Wonthaggi Coalfield, Victoria'. *Proceedings*, No 134. Melbourne: Australian Institute of Mining and Metallurgy.
Hagan, J. 1968. 'Writing Australian Trade Union history'. *Labour History* 14 (May).
Reeves, A. 1986. 'Damned Scotsmen!: British migrants and the Australian coal industry, 1919–1949'. *Common Cause: Essays in Australian and New Zealand Labour History*, edited by E Fry. Sydney: Allen and Unwin.
Saville, J. 1974. 'Class struggle and the industrial working class'. *Socialist Register 1974*. London: Merlin.
Turner, J A H. 1974. 'My long march'. *Overland* 59.

Books and pamphlets

Abrahamson and Son. 1915. *North Park Estate, Wonthaggi*. Wonthaggi: Abrahamson and Son.
Allen, V L. 1966. *Militant Trade Unionism*. London: Merlin Press.
Arnot, R P. 1955. *A History of the Scottish Miners*. London: Lawrence and Wishart.
Australian Coal and Shale Employees Association. 1915. *Rules of the Australasian Coal and Shale Employees Association*. Sydney: ACSEA.
Australian Coal and Shale Employees Federation [Miners Federation]. 1929. *A Review of the Coal Question Preceding and During the Lockout on the Northern Fields, 1929*. Sydney: ACSEF.
Australian Coal and Shale Employees Federation [Miners Federation]. 1931. *Review of the Position of the Australian Coal and Shale Employees' Federation and the Findings of the Committee Appointed by the Federation*. Sydney: ACSEF.
Australian Coal and Shale Employees Federation [Miners Federation]. 1934. *Minutes of the Half-Yearly Central Council Meeting of the Australian Coal and Shale Employees' Federation, August 1934*. Sydney: ACSEF.
Australian Coal and Shale Employees Federation [Miners Federation]. 1934. *The Case for the Union*. Wonthaggi: ACSEF Victorian District.
Australian Coal and Shale Employees Federation [Miners Federation]. 1937. *Six Hours a Day for Mineworkers*. Wonthaggi: ACSEF Victorian District.
Australian Coal and Shale Employees Federation [Miners Federation]. 1937. *There Is a Place for You on the Broad Committee*. Wonthaggi: ACSEF Victorian District.
Australian Coal and Shale Employees Federation [Miners Federation]. 1939. *Interim Award*. Wonthaggi: ACSEF Victorian District.
Australian Coal and Shale Employees Federation [Miners Federation]. 1940. *Fight for Justice – Help the Mineworkers Win Forty Hours!* Sydney: ACSEF.
Australian Coal and Shale Employees Federation [Miners Federation]. 1941. *Declaration of War Policy*. Wonthaggi: ACSEF.

Australian Coal and Shale Employees Federation [Miners Federation]. 1947. *How to Get More Coal*. Sydney: ACSEF.

Australian Coal and Shale Employees Federation [Miners Federation]. 1949. *Why a Coal Strike Looms*. Brisbane: ACSEF.

Australian Communist Party, Wonthaggi District Committee. 1944. *Immediate Problems and Post-War Reconstruction for Wonthaggi and District*. Wonthaggi: ACP Wonthaggi District Committee.

Australian Council of Trade Unions. 1944. *The Mineworkers' Future*. Sydney: ACTU.

Australian Labor Party, Victorian Branch. 1946. *Strikes... and Their Provocation*. Melbourne: ALP Victorian Branch.

Australian Labor Party, Victorian Branch. 1949. *Speak Up for Labor and Liberty*. Melbourne: ALP Victorian Branch.

Australian Trade Union Congress. 1916. *Australian Trade Unionism and Conscription*. Melbourne: ATUC.

Blainey, G. 1963. *The Rush That Never Ended*. Melbourne: Melbourne University Press.

Bolton, G C. 1972. *A Fine Country to Starve in*. Perth: University of Western Australia Press.

Bongiorno, F. 1997. *The People's Party: Victorian Labor and the Radical Tradition, 1875–1914*. Melbourne: Melbourne University Press.

Burgmann, V. 1995. *Revolutionary Industrial Unionism: The Industrial Workers of the World in Australia*. Melbourne: Cambridge University Press.

Butlin, S J. 1955. *War Economy 1939–1942*. Adelaide: Australian War Memorial.

Buckley, K D. 1970. *The Amalgamated Engineers in Australia, 1852–1920*. Canberra: Australian National University Press.

Chambers, J; Chambers, L. 1982. *It's On at the Union: The Wonthaggi Miners' Union Theatre, 1935–1978*. Wonthaggi: Wonthaggi Historical Society.

Child, J. 1971. *Unionism and the Australian Labour Movement*. Melbourne: Macmillan.

Childe, V G. 1921. *How Labor Governs*. London: Labour Publishing Co.

Coghlan, J (ed.). 1979. *The State Coal Mine and Wonthaggi, 1909–1968*. Wonthaggi: Clancy and Co.

Communist International. 1935. *Seventh Congress of the Communist International, July–August 1935*. Sydney: Modern Publishers.

Communist Party of Australia, Wonthaggi Branch. 1937. *May Day: Workers of All Lands Unite*. Wonthaggi: CPA.

Communist Party of Australia, Wonthaggi Branch [J D Blake]. 1949. *The Great Coal Strike of 1949*. Sydney: CPA.

Cooksey, R (ed.). 1970. *The Great Depression in Australia*. Canberra: ASSLH.

Crisp, L F. 1955. *The Australian Federal Labor Party, 1901–1951*. Melbourne: Longmans, Green and Co.

Davidson, A. 1969. *A History of the Communist Party of Australia*. Stanford: Hoover Institution Press.

Day, D. 2001. *Ben Chifley*. Sydney: Harper Collins.

Deery, P. 1978. *Labour in Conflict: The 1949 Coal Strike*. Canberra: ASSLH.

Dimmack, M. 1974. *Noel Counihan*. Melbourne: Melbourne University Press.

Fahey, C. 1987. *The Wonthaggi State Coal Mine*. Wonthaggi: Department of Conservation and Lands and the State Coal Mine Preservation Committee.

Farrell, F. 1981. *International Socialism and Australian Labour*. Sydney: Hale and Iremonger.

Fox, L. 1942. *Coal for the Engines of War!* Sydney: Modern Publishers.

Fry, E. 1986. *Common Cause: Essays in Australian and New Zealand Labour History*.

Sydney: Allen and Unwin.
Gibson, R. 1966. *My Years in the Communist Party*. Melbourne: International Bookshop.
Gollan R. 1963. *The Coalminers of New South Wales: A History of the Union*. Melbourne: Melbourne University Press.
Gollan R. 1960. *Radical and Working-Class Politics*. Melbourne: Melbourne University Press.
Gollan R. 1975. *Revolutionaries and Reformists*. Canberra: Australian National University Press.
Gorman, J. 1973. *Banner Bright*. London: Allen Lane.
Harris, J. 1970. *The Bitter Fight*. St Lucia: University of Queensland Press.
Hobsbawm, E. 1964. *Labouring Men*. London: Weidenfeld and Nicholson.
Hobsbawm, E. 1997. *On History*. London: Weidenfeld and Nicholson.
Hobsbawm, E. 1998. *Uncommon People*. New York: The New Press.
Hobsbawm, E. 2002. *Interesting Times: A Twentieth Century Life*. London: Allen Lane.
Iremonger, J; Merritt, J; Osborne, G (eds). 1973. *Strikes: Studies in Twentieth Century Australian Social History*. Sydney: Angus and Robertson.
Jauncey, L G. 1935. *The Story of Conscription in Australia*. London: Allen and Unwin.
Louis, L. 1968. *Trade Unions and the Depression: A Study in Victoria*. Canberra: Australian National University Press.
Macartney, G. 1894. *The Victorian Coal Consumers' and Investors' Guide*. Melbourne: Troedel and Co.
MacIntyre, A J; MacIntyre, J J. 1944. *Country Towns of Victoria: A Social Survey*. Melbourne: Melbourne University Press.
Macintyre, S. 1980. *Little Moscows*. London: Croom Helm.
Macintyre, S. 1998. *The Reds: The Communist Party of Australia from Origins to Illegality*. Sydney: Allen and Unwin.
Mauldon, F R E. 1929. *The Economics of Australian Coal*. Melbourne: Melbourne University Press.
Murphy, D J (ed.). 1975. *Labor in Politics: The State Labor Parties in Australia, 1880–1920*. St Lucia: University of Queensland Press.
O'Farrell, P J. 1964. *Harry Holland: Militant Socialist*. Canberra: Australian National University Press.
Orr, W; Nelson, C. 1935. *Coal: The Struggle of the Mineworkers*. Sydney: ACSEF.
Orr, W. 1935. *Mechanisation: Threatened Catastrophe for Coalfields*. Sydney: ACSEF.
Orr, W. 1937. *Coal Facts: The Miners' Case for a New Agreement*. Sydney: ACSEF.
Power, F D. 1910. *Powlett Coalfield and Coal History in Victoria*. Melbourne: Rae Bros.
Power, F D. 1912. *Coalfields of Australia*. Melbourne: Critchley Parker.
Powlett Express. 1933. *Speeches by Mr W. G. McKenzie, M. L. A. and the Hon. R. G. Menzies, M. L. A. at Wonthaggi relative to the State Coal Mine, Wonthaggi*. Wonthaggi: *Powlett Express*.
Powlett Express. 1934. *Powlett Express Anniversary Souvenir*. Wonthaggi: *Powlett Express*.
Quilford, A. 1977. *The State Mine: A Pictorial History of the Powlett Coalfields*. Wonthaggi: The Gumnut Press.
Ross, E. 1949. *The Coal Front*. Sydney: Miners Federation.
Ross, E. 1970. *A History of the Miners Federation of Australia*. Sydney: Miners Federation.
Schedevin, C B. 1970. *Australia and the Great Depression: A Study of Economic Development and Policy in the 1920s and 1930s*. Sydney: Sydney University Press.
Shaw, A G L; Bruns, G R. 1947. *The Australian Coal Industry*. Melbourne: Melbourne University Press.
Sheridan, T. 1975. *Mindful Militants: The Amalgamated Engineering Union in Australia,*

1920–1972. Melbourne: Cambridge University Press.

Shire of Korumburra. 1920. *The Land of the Lyrebird: The Story of Early Settlement in the Great Forest of South Gippsland*. Korumburra: Shire of Korumburra.

Sleeman, J. 2008. *The State Coal Mine at Wonthaggi*. Wonthaggi: Wonthaggi and District Historical Society.

Smith, B. 1993. *Noel Counihan: Artist and Revolutionary*. Melbourne: Oxford University Press.

Smith, R. 1981. *Noel Counihan Prints, 1931–1981*, Sydney: Hale and Iremonger.

Stevens, J. 1987. *Taking the Revolution Home: Work among Women in the Communist Party of Australia, 1920–1845*. Melbourne: Sybylla Co-operative Press.

Turner, L A H. 1965. *Industrial Labour and Politics: The Dynamics of the Labour Movement in Eastern Australia, 1907–1921*. Canberra: Australian National University Press.

Victorian Railway Commissioners. 1933. *Victoria's Greatest Asset: True Facts and Figures*. Melbourne: Railway Commissioners.

Victorian Railway Commissioners. 1934. *Wonthaggi Coal Mine Dispute*. Melbourne: Railway Commissioners.

Wonthaggi Co-operative Society (WCS). 1922 to 1938. *Half-Yearly Reports and Balance Sheets*, Wonthaggi: WCS.

Wonthaggi Jubilee Celebration Committee. 1960. *Wonthaggi Jubilee Souvenir*. Wonthaggi: Wonthaggi Jubilee Celebration Committee.

Wonthaggi Trades and Labour Council. 1945. *Wonthaggi and District Post-War Reconstruction: Plan for Happiness, Peace and Stability*. Wonthaggi: Wonthaggi TLC.

Workers International Industrial Union, Wonthaggi Local. 1918. *Workers of Wonthaggi!* Wonthaggi: WIIU.

Government publications

Victorian Government. 1909–10 to 1967–68. Annual Reports of the General Manager of the State Coal Mine. Melbourne: Victorian Government.

Victorian Government [Lee, R]. 1934. 'Report on the Victorian State Coal Mine'. Melbourne: Victorian Government.

Victorian Government. Victorian Parliamentary Papers. 1906. 'Royal Commission on the Coal Industry, Labour in Mines, Health of Miners, Settlement of Disputes, Etc.'. Melbourne: Victorian Government.

Victorian Government. Victorian Parliamentary Papers. 1907. 'Report from the Select Committee on the Mining Industry'. Melbourne: Victorian Government.

Victorian Government. Victorian Parliamentary Papers. 1937. 'Interim report of the Royal Commission on certain matters relating to the State Coal Mine, Wonthaggi'. Melbourne: Victorian Government.

Victorian Government. Victorian Parliamentary Papers. 1937. 'Second report of the Royal Commission on certain matters relating to the State Coal Mine, Wonthaggi'. Melbourne: Victorian Government.

Newspapers and journals

Age, Melbourne.
Argus, Melbourne.
Australasian, Melbourne, 1914.
Common Cause, Sydney, 1920–1924; 1934–1968.
Communist Review, Sydney, 1934–1950.
Criterion, Wonthaggi, 1910–1913.

Miners Union Bulletin, Wonthaggi, 1949–1968 (intermittent).
Miners' Voice, Wonthaggi, 1934.
Mining and Engineering Review, Melbourne, 1910–1914.
One Big Union Herald, Melbourne, 1919–1923.
Powlett Express, Wonthaggi, 1910–1950.
Red Leader, Sydney, 1933–1935.
Sentinel, Wonthaggi, 1910–1950.
Socialist, Melbourne, 1910–1914.
Sprag, Wonthaggi, 1933–1934.
Sun News-Pictorial.
Union Voice, 1934–1968 (intermittent after 1950).

Index

accommodation, lack of   3, 66, 126
Age   31, 32, 40, 51, 90, 141
Amalgamated Engineers Union (AEU)   114, 145, 147, 148
Asquith, Arthur   34, 59, 63, 69, 71, 74
Australasian Coal Miners Association (ACMA)   14–15
   see also Miners Federation of Australia
Australia at War exhibition   130
Australian Coal and Shale Employees Federation (ACSEF)   xi, 15, 151
Australian Labor Party (ALP)   xvi, 16–18, 23–25, 37, 38, 39, 50, 54–55, 66, 67, 74, 84, 103, 105–106, 108–111, 114, 117, 118, 132, 133, 137–139, 146–148
Australian Railways Union (ARU)   114
Australian Socialist Party   27
Australian Timber Workers Union   103
Blake, J D   106
Bloustein, Harry   15
blue pigs (shanty stills)   3
British immigrants, *see under* immigration
Broken Hill   xv, 29, 73
Broome, George   2, 9, 12–13, 19, 31, 33, 36, 39, 57, 60
Brown, James   115, 147
Bruce, Stanley Melbourne   55, 67
Brydon brothers   42
Burley, Tom   14, 15
Burns, Jack   17
Byrne, Jim   151, 152
Cain, John (Sir)   55, 146
   Cain government   133, 137–138
Caledonian Hotel   16, 18, 44, 72
Cameron Street   100
Campbell, Merv   151, 152
Canberra Code   100
Cape Paterson   xiv
Case for the Union   70, 71
cavilling system   12, 13, 42, 49, 65, 100, 120
Central Victoria   8
Chambers, Agnes   73, 105, 112, 113
Chambers, Jim   49, 105, 112
Church Hill   100
Citizens Defence League   22
clippers   7
coal carters   75
Coal Creek Company   xiv, xv

Coal Facts   82, 85
coal industry   xi, xiii, xiv, xvii, 9, 31, 33, 51, 79, 82, 86, 104, 105, 117, 120, 130, 132, 133, 148, 149, 150
Coal Mines Regulation Act 1909   3
Common Cause   65, 85, 90, 102, 120, 121, 123, 132
communism   56, 106, 139
Communist Party of Australia (CPA)   61, 106, 108, 110, 111, 113, 118–119, 124, 132
communists   50, 61–64, 66–68, 70, 84, 103, 105–115, 117–121, 124, 130, 131–132, 138–141, 144, 147–148
Conciliation and Arbitration Act 1904   xv
Connelly, John   59
contract system   26
conventions, *see under* Miners Federation of Australia
Co-operative, *see* Wonthaggi Co-operative Society
Cosgrove, Mick   3
Counihan, Noel   128–130
Criterion   1, 2, 12, 14, 15, 16, 17, 18, 21, 22
Crystal Palace   41
Currie, Tom   71, 118
Curtin, John   120, 122, 123
Dalyston   1, 91, 138
Davidson Royal Commission   123, 133
Davies, Bert   24, 29–30
Doig, Agnes   134, 143
Doig, Wattie   87, 88, 143
Dowling, W J   33, 34, 54, 61
Dudley Area   34, 35, 69, 83
Durham family   38
East Wonthaggi Australian Rules Football Club   50
Eastern Area   34, 83, 126, 149
Easton, William   36, 39, 59, 114, 130
eight-hour day campaign   25, 104
Eight Hours Day   13, 14, 20, 151
Eildon Weir   28
electrical coal cutters   2
explosion, 1937 (Twenty Shaft)   90–101
Fairless, Ted   90
Farmer, Alan   148–149
First and Second Rounds of the Log   85, 102, 103–104, 106, 109, 115, 116, 117, 119
First World War   14, 18, 21, 22-26, 30, 31,

32, 34, 35, 59, 67, 113
Fitzpatrick, Brian   127, 130
Flinders   55, 67
Foster, Joe   49, 71, 143
Foster, Meg   134, 135, 136
Friends of the Soviet Union (FOSU)   140
Gibson, Ralph   68
Gippsland mines   xiii, 87
Gippsland region   xiii, xiv, xvii, 8, 9, 19, 56, 77, 114, 117, 125, 149, 150, 152
goldminers   xvii, 1, 8, 9
Goldsmith, Jack   11, 17–18, 23, 27, 61, 63, 119
Graham Street   17, 41, 43, 44, 72, 79, 100, 125
Great Depression, the   xi, 48, 51, 54–57, 61, 64, 67, 72, 77, 79, 80, 82–84, 87, 107, 116, 118, 119, 138
Grieve, Nancy   49
Hagelthorn Street   50, 100
Hamilton, Bob 'Hammie'   49, 76, 87, 118, 131, 132, 142, 147, 148
Hamilton, Elsie   134, 135
Harmers Haven   42
Hicksborough   34, 118
Hogan, E J   54, 55
Holland, Bert   8
Hollole, Keith   91
Holloway, E J 'Ted'   55
Hughes, Billy   24, 26
Hunter, Stanley   xvi, xvii, 1
Hyland, Herbert   127
immigration:
    British   31, 35, 42, 63, 105, 112, 119
    Italian   50, 117
Imperial Peace Conference   26
industrial dispute   xiv, xvi, 12–13, 14, 25, 32–34, 37, 50, 54, 59–64, 66–78, 83, 84–85, 87–89, 100, 103–105, 107, 114, 115, 119–123, 139–145, 148, 151
industrial militancy   xv, 21, 26, 27, 30, 32, 42, 59, 90, 120, 122, 137, 146–148
Industrial Workers of the World (IWW) ('Wobblies')   26, 27
Inverloch   xvii, 42, 118
Irvine, William (Sir)   23
Irving, Jock   118
Italian community   50–51, 117
Johnston, Bob   133
Johnston, Thomas   90
Jones, Jimmy   69
Jumbunna   xiv, 8, 77, 87, 112
Jumbunna Coal Company   xiv
Kent-Hughes, Wilfred   144, 146
Kilcunda   xiv, 55, 77, 87, 103, 117, 138
Kirrak Basin   39, 42, 83, 114, 116–117, 126, 149

Korumburra   xiv, 8, 14, 30, 87, 88, 102, 110, 112, 127, 138
League of Young Democrats   111
Lee, Robert   77, 78
Lees, George   90
leisure activities   40, 41, 50, 61
lockout:
    Gippsland, 1903–04   xiv
    Northern District, 1929–30   51–52, 57
Log campaign, *see* First and Second Rounds of the Log
Log of Claims   103, 132, 141
Long, J J   32
Lovegrove, 'Dinny'   114
Lysaght steel mill   112
Maitland (NSW) coal   53
Mann, Tom   17
Mauldon, F R E   79, 82
May Day demonstrations   74, 100, 134, 135, 136, 137, 151
McBride, Peter   xvi, xvii
McBride Street   18, 38, 79
McBride Tunnel   24, 35, 83
McColl, J H   xiii
McCoy, Frederick   xiv
McKenzie, John   144, 147
McKenzie, W G 'Bill'   37–40, 51, 54–59, 64, 83, 84, 107–110, 114–115, 117–120, 130, 133, 137–139, 145
McLean, John   105
McLeish, John   57–66, 68–71, 77, 83, 90, 91, 93, 100–101, 103, 116–117, 121, 123, 133
McMahon, Matthew   8, 10, 11, 16, 23
McVicars, John 'Jack'   3, 11, 30, 34, 54, 57, 59, 61, 63, 66, 69, 71, 74, 76, 80, 87, 131, 150
mechanical coal-cutters   19, 86, 102
mechanisation   79, 82, 86, 102, 109, 121
Melbourne Trades Hall Council   55, 74, 75, 106, 114, 151
Menzies, Robert G   64, 75, 77, 118, 140, 148
military conscription   22–26
Miller-Cyclone Forgings   130
Miners Federation of Australia   25–30, 34, 39–40, 51, 54–58, 60–62, 65, 67, 69–75, 77–78, 79–87, 90, 100–105, 108–112, 114–115, 119–124, 130, 131–133, 137, 139–142, 145, 146–150
    1938 National Convention   103
    1943 National Convention   132
Miners Federation of Great Britain   67
Miners' Voice   67, 74, 85
Mineworkers' Future, The   127
Mining and Engineering Review   19
Minority Movement (MM)   61–64, 66–70, 71, 73–74, 85, 105, 106
Mrs Connelly's Orchestra   41

Murphy, Frank   8, 10, 11, 14, 16
National Coal Tribunal   32
Nationalist Party   24
Nelson, Archie   17, 23
Nelson, Charlie   82, 85, 87
New Zealand mines   2, 14, 21
Newcastle coal   xvi
North Wonthaggi   34, 118
Northcote Town Hall   73
Northern Area   87, 126
OBU Victorian Organising Committee   27
Official OBU   27–29
One Big Union (OBU)   26–31
Opie, Alan   118, 121
Orr, William (Bill)   66, 69, 73–75, 77, 82, 85, 116, 131
Outtrim   xiv, 8, 9, 12, 13
Outtrim, F L   xiii
Outtrim-Howitt Company   xiv
Pakenham   75
Parliamentary Select Committee Inquiries   xv, xvi
Peacock, Alexander (Sir)   22
Peter Bowling Strike, *see* strike, Peter Bowling
pit committees   70–71, 75, 77, 121, 123
pit ponies   7, 63, 65, 70
Political Labor League (PLL)   15
Political Rights Committee   118
Powell, Harriet   15
Powlett Express   3, 9, 13, 14, 36, 40, 55, 56, 57, 60, 75, 90, 102, 133, 138, 140
Powlett Hotel   44, 57, 72
Powlett Plains   xvii
Powlett River seams   xvi, 4, 7, 10, 21, 129
Prendergast, George   xvi
preferential employment   49
Premier's Plan   55, 105
Propaganda Committee, Wonthaggi mineworkers'   73, 143
Public Assistance Committee   65
public coal sales   19, 21, 36, 37, 39, 53, 64, 108
Rankine, Bill   50
Red Fed (New Zealand Federation of Labour)   14
Red Leader   68, 69
remote-control endless-rope haulages   2
Richmond Brewery   57
Rodgers, John   140
Ross, Edgar   15, 26, 132
Royal Commissions   xiii, xiv, xvi, 100–104, 123, 133
Russell, 'Bob'   54, 61, 63, 66, 71, 143
Rutherglen   xvii, 8
Sale of Coal to the Public League   19
Samson, Sam   132
Scott, J B (Jim)   28–30

Second World War   114, 116–121, 123, 125, 126, 130
Semple, Bob   14
Sentinel   12, 13, 15, 18, 23, 26, 36, 37, 38, 39, 52, 54, 64, 84, 90, 100, 102, 107, 108, 109, 110, 113, 115, 126, 127, 130, 138, 140, 144, 147
Sharkey, Lance   148
Short, John   40, 71
'Shovelman'   2, 3
soccer league   49, 56
South Dudley   34, 138
South Gippsland Miners Association   xv, 30
Speirs, Ron   90
Sprag   67, 71, 74
State Mine Tribunal   32
Station Area   18, 34, 43
Stirton family   49
   Stirton, W (Bill)   49, 50, 53, 71, 74, 75, 77, 87, 105, 114, 143
   Stirton, Nancy   134, 135, 136
Stockton Mine (NZ)   2
Stout, J V   114
Strzlecki Ranges   xvii, 8
strike action:
   1932 strike (State Coal Mine)   59–60
   1934 strike (State Coal Mine)   xi, 50, 59–64, 66–78, 79, 83, 84–85, 108, 111, 112
      Entertainment and Relief committees   50, 72, 75
   1935 shaft-sinkers strike (State Coal Mine)   84
   1949 strike (State Coal Mine)   139–146
   Peter Bowling Strike   xvi, 3
   Sunbeam mine strike (1937)   87–89
Sunbeam mine   87–89
Sunraysia   107
Sydney Trades and Labour Council   62
Taberner's Wonthaggi Hotel   18, 30, 48
Third Australian Peace Conference   26
Turner, Ian   26–27
Union Theatre   41, 43, 45, 46, 56, 72, 73, 79, 90, 140, 143, 144
Union Voice   85, 86, 90, 100, 101, 102, 108, 110
Victorian Coal Miners Association (VCMA)   xiv, xv–xvi, 8–9, 12–16, 30, 59, 151
Victorian Coal Tribunal   62, 69, 109, 141
Victorian Colliery Owners and Employees Federation   xv
Victorian Council Against War and Fascism (VCAWF)   106, 111, 113
Victorian Miners Tribunal   38
Victorian Mines Department   xvi, 7
Victorian Parliament   49
Victorian Railway Commissioners   7, 19,

21, 32, 33, 36, 39, 51, 53, 55, 56, 58, 63, 70, 73, 74, 103, 104, 108
Victorian Railways   xiii, xiv, xvi, 7, 35, 36, 75
Victorian Socialist Party (VSP)   17, 23
Vistarini, Ed   71
Walhalla   8
Webb, Harry   10, 130
Webb, J W   16
Webb, Paddy   38
Wells, Harold   123, 124, 131
West Fife   42, 49, 50
Western Area   5, 39, 42, 83, 101, 102, 116, 126
wheelers   xv, 7, 13, 23, 49, 59, 65, 129
Williams, Idris   50, 60, 61–63, 67, 69, 71, 74, 78, 87, 90, 102, 106, 107, 114, 115, 118, 119, 123, 127, 130, 131, 141, 142, 144, 145, 146
Willis, A C   27
Wilson, Arthur   150
Winneke, Justice   62, 86
'Wobblies', *see* Industrial Workers of the World
Wonthaggi Advancement and Defence League   108
Wonthaggi and District Postwar Reconstruction Plan for Happiness, Peace and Stability (the Plan)   124–127
Wonthaggi and District Vigilance Committee   149–151

Wonthaggi Borough Council   23, 36, 56, 61, 103, 109, 126, 144
Wonthaggi Brass Band   50
Wonthaggi Citizens Band   88, 90, 100, 103
    see also Wonthaggi Union Band
Wonthaggi Coal Mine Dispute   70
Wonthaggi Co-operative Society   40, 56, 72, 102, 112, 125
    Co-operative Dental Clinic   40, 125
    Co-operative Dispensary   17, 79
    Co-operative Store   17, 38, 40, 75, 79
    Friendly Society   40
Wonthaggi Miners Women's Auxiliary   100, 103, 111–113, 134, 135, 140
Wonthaggi RSL   113, 140
Wonthaggi Social Club   50, 111
Wonthaggi Unemployed Workers Union (WUWU)   56, 65, 108
Wonthaggi Union Band   20, 22, 37, 41, 48, 61, 87, 88
    see also Wonthaggi Citizens Band
Wonthaggi Workingmen's Club   17, 40, 50, 79
Workers Industrial Union of Australia (WIUA)   27, 28
Workers International Industrial Union (WIIU)   27–30
Workers' Sports Federation   111
Wren, John   16
Yardley, Ern   8, 9

In the narrow seam   12/50   Wace Lounsha '47